Raising Pigs

Beginners Guide to Raising Healthy and Happy Pigs on a Small Homestead

By: Irene Mills

Table of Contents

Introduction

Pigs on a Spring Pasture at Smith Meadows Farm[1]

Pigs can be an incredible asset to a small homestead. Besides meat, they can have functional benefits on the land. In addition, they are intelligent and can be relational and even affectionate with the proper training.

This book will cover pig basics (food, shelter, set up, breeds, etc.) as well as discuss the uses of pigs on your homestead, how to train them to be easier to handle, breeding, and butchering them for meat.

This book is written as an essential guide for beginner pig keepers on a small homestead. We focus on the smallest of the heritage breeds, American Guinea Hogs and Kunekune pigs, assuming small herds of 2-8 pigs. If you choose larger breeds to work with on your homestead, then this book will provide a solid foundation for keeping pigs on your land, but you will need to find the information on the breed you choose specifically. If you have the acreage to have 10+ pigs, this book will be an excellent foundation for you, but you will need to supplement with resources and information about commercial pigs.

FAQ

What's the difference between a pig, a hog, and a boar?

Good question. This can be confusing because, in common usage, the words "pig" and "hog" are often synonymous.

Technically,
- A pig is a domestic breed of less than 120 pounds.
- A hog is a domestic breed more than 120 pounds.
- Boars are wild or feral. Wild Boars usually weigh 400-600 pounds, and have formidable tempers translated and expressed through their tusks. The weight of feral boars depends on the breed of pig that was released and became feral. In the United States and other countries, they can be feral and invasive. Wild and feral boars do an incredible amount of damage to ecosystems.

For the purposes of this book, we will be using the word "pig" generically to refer to domestic swine, per common usage.

Am I allowed to have pigs on my homestead?

Check with your municipality regarding the zoning laws on your property. Even if you are allowed to have pigs, there may be restrictions on placement beside waterways and property lines, the number of pigs you can have, etc. Get the details before you make your plans.

Should I have pigs?

Another good question. This book will give you an idea of what is involved in keeping pigs. The following FAQs will also help you decide.

How many pigs can I have? How much space do they need?

You can have 2 pigs on a ½ acre for a permanent enclosure.
A maximum of 10 pigs can live on an acre if you are doing brush clearing.[2]

You need to have at least 2 pigs; they are herd animals and not happy alone.

Do pigs need mud and water?

Yes, they need mud and water, but you don't need a pond and a permanent mud space. Contrary to popular belief, they don't want to be in mud all the time; in fact, it is not suitable for them. They enjoy mud wallowing when it is hot, but they will get hoof rot if kept in mud constantly. They enjoy a pond when it is hot, they will love it, but it is not a necessity – a mud puddle will suffice.

Do pigs smell?

Pig urine and poop scents are strong and repellent to humans. When it builds up in one enclosed space, it is beyond intense. Pigs themselves don't smell, but the stench of their urine and feces will linger in a space after they have been moved out of that enclosure entirely.

Why do some pigs have nose rings?

Some pig farmers put nose rings on their pigs to keep them from rooting up the ground. Pigs love to eat roots and will dig down deep with their snouts and turn over the soil. This can be a pro or a con. We will discuss this further in the functions of pigs on a homestead. You may decide to have some pigs that graze the plants on top, and some who turn up soil and compost, depending on how many pigs and spaces you have. Most homesteaders prefer to "let pigs be pigs" and choose Kunekunes if they want pigs who are less inclined to root deeply.

Do domestic pigs have tusks?

Yes. All breeds have tusks - domestic, feral, and wild. They are a pig's defense against predators. Some owners trim or remove tusks. More information follows in the health maintenance section.

Can humans get sick from pigs?

Yes. Humans can contract "swine flu." It is transmitted from pigs the same way other cases of flu transmit between us: through droplets after coughing, sneezing, or contact with something that has the virus on it.

You are at most risk if you go to county fairs and walk into a pig barn, but you are also at risk of exposure when you visit breeders and other farms.

The CDC has published a helpful article and guidelines regarding swine flu risk.[3] It is oriented toward those showing their pigs at fairs, but the advice applies to any pig owner who may visit other farms.

Primary preventions listed by the CDC are:

- Wash your hands with soap and water or hand sanitizer before and after going into a pig barn or enclosure.
- Do not take toys, pacifiers, cups, bottles, strollers, or children into pig enclosures with you unless they are of age to be learning to care for pigs and can be trusted not to touch their mouth or eyes before washing their hands when they are out of the pig enclosure.
- Do not eat, drink, smoke, or put anything in your mouth while in the pig enclosure or before you wash your hands after you leave.

Homestead Pig Vocabulary

- AI – artificial insemination
- Anthelmintic – A drug treatment for parasitic worms. Ivermectin is commonly used for this purpose
- Blind teat – A teat that is unable to be used. These are usually removed so that piglets don't suck on a dry teat
- Body capacity – Calculations are used to determine the potential quality of the carcass
- Colostrum – the nutrient-rich first milk from the sow after birth. The piglets feed on the colostrum for a few weeks as they build their strength and immune systems
- Culling – removing any pigs that are not desirable for breeding
- Drove – a herd or group
- Farrow – a verb - to give birth to pigs
- "Farrow to Finish" - from birth to butchering
- Feeder pig – This term is used in two ways: 1) Any pig being raised for pork. 2) More specifically, the term "feeder pig" can be used as a term for pigs 6-10 weeks old born on one farm, then purchased by another to raise to butchering
- Piglet – Newborn pig to 6 weeks
- Weaner - 6-8 weeks old
- Shoats – A young pig that has recently been weaned and is not yet full adult size

- Gilt – A female that has not had piglets
- Sow – A female that has had piglets
- Boar- male, intact
- Barrow – male, castrated
- Pigsty – an enclosure for pigs
- Creep Feeding – A protocol used to transition a piglet from mother's milk to solid feed[4]
- Wallow – A depression in the soil that pigs make for mud and dirt baths

Deciding Whether You Want Pigs

Pros and Cons of Pigs on a Homestead

Pros

The correct number of pigs of an appropriately selected breed can be incredibly useful on a homestead. The uses we will cover in this book are:

- Meat for your family
- Artisan meat sold "whole hog" (we will discuss how to do this legally)
- Compost machines
- Forest farming
- Brush clearing
- Rototillers
- Pond sealing
- Truffle hunting
- Showing

Rotated properly on a piece of landscape, they become your partners in farming, landscape maintenance, and compost disposal. If you have children, pigs can be a part of a 4H project or simply your own livestock training for your kids. In addition, the heritage breeds described in this book provide meat that is incredible.

Cons

Consider these points carefully, as you do not want to take on pigs if your land or lifestyle is not suited to them, or if you are not ready to accommodate them.

- Pigs can be extremely useful, but they also run the risk of being terribly destructive. You need to pay attention to what they are foraging and be prepared to move them in time before they start doing damage. If you do not rotate them, the area will become a muddy rototilled area.

- They need secure fencing. Pigs are super smart, and if bored or not getting enough to eat, they will seek ways of getting out. They can be excellent escape artists. If you are rotating them, ensure that the electric fence is set up securely and maintained. They *will* test it. We discuss training them to respect the fence in Part Two: Getting Set Up.
- As with any livestock, you'll get more flies. Having a flock of hens in the pig area will make a big impact on the fly situation. I tested this over the course of 4 years.
    - Year 1: no chickens, lots of flies, moths, and beetles on our land.
    - Year 2: had chickens, radical reduction in flies, beetles, and moths.
    - Year 3: no chickens, fly population back up, beetles not as numerous as they were year 1, but more than year 2.
    - Year 4: had chickens, summer fly population was low, other insects back down.
  I decided to always keep chickens around.

- You will probably want to invest time in training them. Feeding, managing, and slaughtering are all much easier if you put some time into training. We go into detail about training in Part Three.

You can see from the "cons" listed above that these are not negative points - they are *commitments* that you need to consider to avoid damage. The list above may not be cons in your situation. Starting small with just a couple of pigs for a season is a wonderful way to get your systems set up and get a sense of what it is like to keep them. You can always slaughter and not replace them if you decide this isn't for you.

Another way of working with pigs is to keep them around while they have jobs, slaughter, or sell them, then wait to get more. You have the option to grow into the practice of pig keeping and then let it go if their uses are finished, and your freezer is filled. We have done this on our homestead; the pigs were around for a few years living in various spaces that needed clearing. Then they were out of a job and were just living in the stockyard, making a mess and being bored. We had a break from raising pigs until we were ready to set up our pond for sealing. Since then, we've set up fence posts for the electric fence in a brushy forested area.

Part One: Breeds

When most people think of pigs, they imagine the super-sized pink pigs commonly used in industrial mass production. Although these pigs are not recommended for homesteads, a homesteader must understand the criteria to choose an appropriate breed from those suited to small homesteads.

The breeds used on homesteads are called "heritage breeds." The modern industrial breeds have been bred to live inside in a tightly confined space. There are three main reasons why the super-sized pink pigs are not recommended for homesteads: 1) They can weigh 600-800 lbs.; 2) They don't have any hair, so they get sunburned easily and do not have winter protection either. And 3) They are lethargic and not nearly as relational as heritage pigs.

Heritage breeds have been bred outdoors for hundreds of years. They will not get sunburned as easily as the (nearly) hairless, pink-skinned pigs. They do need shade and shelter, but their hair protects them from both sun and cold. I live in Montana, and my American Guinea Hogs have done fine in the winters with deep snow as well as the sunny days with temps in the 90s in July and August. A simple shelter with straw bedding keeps them happy in the winter. In the summer, given a shady place to hang out, they will keep themselves from getting sunburned. When it was in the mid 90's and triple digits, I would hose them off once or twice a day (they love it).

There are quite a few heritage breeds to choose from. This footnote will list several articles that outline them.[5]

We will focus on two breeds which are on the most petite side. Both breeds are favorites of permaculture practitioners and those with small homesteads. If you think you want to raise larger pigs, either of these breeds would help you learn what to expect and how to manage them on your landscape while being less demanding of your space and easier to handle than the larger breeds.

American Guinea Hogs (aka AGHs)

American Guinea Hog Association

"Life Cycle of American Guinea Hogs"[6]

History:

The history of this hog is murky. It is most likely that AGHs came to Europe and the Americas from West Africa, but there has been evidence of DNA links with the "Improved Essex Hog" in the UK. It's possible that they took a step into the UK prior to coming to the US. In her book, Saving the Guinea Hogs: Recovery of an American Homestead Breed, Carol Payne goes into detail about the breed's origins.7

After a time of being popular for homesteads and small farms in the U.S., the American Guinea Hog was almost extinct. After WWII, people started moving away from farming and homesteads and preferred to purchase their meat rather than growing their own or be market farmers. This migration from farm to urban living led to the decline of the smaller pigs used in homesteading and the rise of breeding for factory-farmed pigs that gained weight fast, had extremely low energy, could be kept inside and confined to small spaces.

Payne goes into the history of the recovery of the American Guinea Hog (AGH). Her efforts have played a big part in their restoration and the continued standards for the breed.

Weight:

150-300 pounds

Height:

22-27"

Length:

46-56"

Temperament:

American Guinea Hogs are known to be docile and non-aggressive. Other swine breeds can be dangerous to chickens, cats, dogs, and children as well as human adults. The AGH is content to let chickens peck around in their enclosure (a great plus to keep flies down). I've heard of several owners who trained their AGHs to come when called, roll over and get a belly or back scratch.

The mothers tolerate handling during farrowing and raising newborn piglets, especially if you have established trust and they are accustomed to being handled. Having said that, all mammals will have protective maternal instincts. Don't jump in the enclosure and grab a piglet. Build trust, be calm, patient, and negotiate.

America Guinea Hogs are also popular with those acquiring pet pigs because they can be house-trained and are so amiable and gentle.

Lifespan:

15-20 years

Usual age at slaughter:

6 months is the earliest to get full-grown adult meat. The American Guinea Hog Association (AGHA) suggests that you strategize your butchering cycle depending on what you want.
- You get about 50% of the weight in meat when you butcher.
  - At 6 months old, they weigh about 60 lbs., so you will get about 30 lbs. of meat.
  - By 12 months old, they will be about 120 lbs., so you will get about 60 lbs. of meat.

o   See more detail in the Butchering section.

Special characteristics:

- American Guinea Hogs LOVE grass. That is why they dig later than other pigs. A grass-fed AGH is healthy and produces excellent meat. This feature also keeps feeding costs down.
- Will start rototilling later than other pigs, so they are easier to manage on a landscape.
- They tend to be excellent mothers. Rejecting piglets is rare. Typically, nature takes its course, and they are low maintenance.
- In most pig breeds, you need to choose between breeding males or castrating them because the testosterone taints the taste of the meat. However, for some reason, AGHs are an exception to this. We have heard this from others and found that our boar produced excellent meat after siring piglets.

Resources

The American Guinea Hog Association is the best place for finding reliable information about care and a directory of registered breeders.
https://guineahogs.org/

Kunekune Pigs

Morning Chores Blog[8]

History:

Kunekune pigs are from New Zealand, but they are not indigenous. We don't know how they got there; the most likely scenario is that trading with the Maoris brought them from Asia or the Polynesian Islands. The Maori people raised them for meat. Kunekunes almost went extinct, but a couple of people gathered 18 of them in the 1970s, formed an association, and now they are on a path of steady recovery. Kunekune pigs are now a well-established breed in the U. S. as well.

Weight:

140-220 lbs.

Height:

24-30"

Length:

approximately 48" long

Temperament:

They are very gentle and trainable. One thing about their history is that the Maori people would let them roam around their camps. This constant exposure and interaction with humans may have resulted in their social nature over time. They are exceptionally social and thrive on a lot of human interaction. They are the best choice for pet pigs. Even if you raise them only for meat and no other uses on your homestead, make sure that you give them interaction and attention.

Lifespan:

15-20 years

Meat:

They produce high-quality meat and are ready for slaughter when they are 14-18 months old.

Special characteristics:

- Kunekunes exhibit a range of mixed colors. If you want a picturesque gathering of small friendly pigs, these are for you.
- Their snout is shorter than the AGH. Kunekunes are grazers. They will not rototill to the extent that the American Guinea Hog will. This is advantageous if you don't want any rototilling, but a disadvantage if you need that function on your land. They prefer to eat grass.
- Kunekunes have "wattles" known as *piri piri* which are flaps of flesh that hang off the jaw. Breeders have tried to figure out what makes the *piri piri* occur in some pigs and not others but have not found answers in lineage or tracking their bred pigs. So until we find a reason, *piri piri* seem to manifest randomly.
- Kunekunes are not escape artists and are not inclined to want to get out of their enclosure. They will respect an electric fence.

Resources:

The American Kunekune Pig Society (AKKPS)
https://americankunekunepigsociety.com/
The AKKPS has a lot of helpful information, including members who are reputable breeders, pigs for sale, "help guides," and more.

Other Homestead Breeds

Of course, there are quite a few other heritage breeds that do well on homesteads. We have highlighted the American Guinea Hog and Kunekune because of their small size, docile nature, and ability to provide various uses on a landscape with less risk of damage. We recommend these as the best breeds for beginners.

As noted, in the footnotes you will find articles that discuss other larger breeds as well. You may decide you are up to starting with one of the larger breeds, or you may have a cycle or two of Kunekune pigs or American Guinea Hogs and then decide that you want to explore another heritage breed.

Whatever breed you choose, do your research into their particular feeding requirements, temperaments, and environments that make them happy and healthy. Make sure that your pigs are a good match for your land and your needs.

Part Two: Getting Set Up

Prior to adding pigs to your homestead, decide precisely how they will be used on your land. Do some calculations to determine how often they can be rotated over your landscape. We recommend starting small. Set yourself up for success by making your first season with 2-3 pigs. Experiment with rotating them through your landscape or keeping them in their pen if you only use them for meat. If you have the space and the need, you can always get more or breed them.

I was pleased with the strategy when we were new to pigs. We purchased three young adult females who had not yet given birth (gilts). We watched their personalities and figured out which ones we wanted to breed. One of them did not have the docile temperament typical in the American Guinea Hogs. She wasn't exceptionally aggressive but certainly not social like the others. She went into the freezer, the others we kept for breeding. Had we introduced a boar when we got all three of them, that troublesome pig would have been bred. Besides temperament, over time you may discover some genetic characteristics that you do not want to breed. Observation of your pigs before breeding is well worth the time.

As you research the breeder where you will purchase your pigs, you will also need to get ready so that your new pigs come home to a great place to settle in and live. An organized and complete setup will also help reduce your tasks as you become accustomed to caring for your pigs daily.

Shelter & Barnyard

Shelter

Your pigs need a shelter inside their enclosure that is dry and sheltered from the wind and drafts. Pigs make a latrine, so they do not soil their bedding. In the winter, a thick bed of dry straw will make comfortable bedding. The only tending it will need will be topping off.

It is critical to use straw, *not* hay, for the bedding. Hay has the seed in it. It is alive and has nutrition. Pigs will eat it, and it will mold and rot. Straw is the chaff that does not have any seed or nutritional value, will not be eaten, and is not prone to mold and rot.

The possible shelters you might use are limited only by your imagination. You can build a wood lean-to with an opening for the pigs to go in and out; add a door so you can shelter them from the elements and get into it when necessary. Even if you are going to move your pigs around your land, you might find it handy to have a simple shelter set up in a permanent stockyard. Our permanent shelter is multi-purpose; the chickens used it as a nesting place, the cows have used it for birthing, and goats have used it in the winter. A permanent shelter that is easy for you to access can be critical if you need to isolate a pig from the rest of the herd or have a vet come in for examination or treatment.

Your local farm supply company will sell permanent pig shelters. However, if you want to build it yourself, here is a site with a DIY plan complete with a detailed list of materials necessary.[9]

For a mobile shelter, we use an upside-down stock tank. We cut a hole in it so that the pigs could get in and out. It worked beautifully and had the advantage of being mobile. Since we move the pigs around the land, it is crucial to have a mobile shelter that is solid enough for Montana winters.

The pigs enjoy a tight, cave-like space, but they need to be able to get in and out and turn around comfortably in there. I really loved this stock tank. It is big enough for three, and it is made of non-toxic recycled material. We cut a hole as a door and laid the straw bedding down. We have had happy, warm piggies in 2-2½ feet of snow, usual temps in the teens and 20s F with drops into single digits and below zero at night and even a few days.

In the spring, we pulled the tank up and distributed the straw around the meadow where the pigs were working. You can also use it as compost or for other barnyard/stockyard uses. As the spring and summer approached, they used it less and less, and it became an excellent place for farrowing. Our new mama sow settled in it happily, and 8 piglets thrived being born and raised for 10 weeks before weaning.

Another option is to park a stock trailer in the enclosure. Make a sturdy ramp and line it with straw. You could even line the walls by stacking straw bales if you are in a freezing environment. I knew a farmer who did this, and he made an entrance by cutting particle board to size with an archway in it and fastened it to the open trailer. If your trailer has doors, then you just need to prop it open enough for them to get in and out and secure it in place.

You can also purchase mobile pig shelters. For example, the shelters shown below would be excellent choices.

This one comes with extra options for a guard rail and doors on both ends to keep pigs inside.

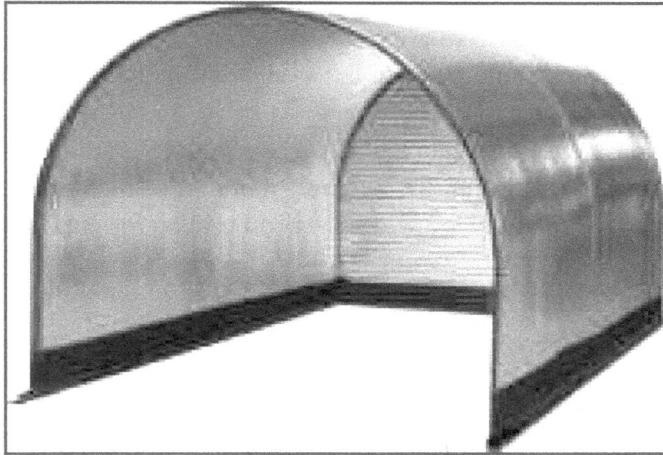

Port-A-Hut, Inc.
http://www.port-a-hut.com/individual-hut.cfm

For farrowing and providing a space for quarantine, the same company offers a farrowing pen option.

Port-A-Hut, Inc.
http://www.port-a-hut.com/farrowing-pens.cfm

Another joy for them in the winter is to put fresh straw out on the snow (or mud) on sunny days. They love to get out and catch some fresh air in the sun during the winter months.

When the weather gets above freezing in the spring, we found that they spent their days outside and only used the shelter in the rain. By summer, there were many nights when they stayed outside under the stars. Clean straw is always a welcome thing for pigs for an outdoor hang out space.

Fencing

If you have a permanent stockyard area, then permanent fencing is worth the investment. However, if you are building with pigs in mind, think about how small they are and how you will keep piglets inside the fence. You may need to add electric fencing to the bottom of their permanent fence till the piglets grow too big to get out.

Electric fences are used for temporary enclosures around homesteads. For example, if you are rotating your pigs around your land and you know they will come back to a particular space, you can install the supporting posts and let them stay there till the next rotation. You can also move the posts along with the fence when you rotate the pigs.

Do your research first to understand the guidelines for purchasing suitable materials and setting up a secure electric fence. These articles will give you an excellent start.[10] As you research, you will notice that electric fencing for pigs is not the same as providing it for brush goats or other animals. Like goats, pigs are highly intelligent, but they are lower to the ground and tend to root. You don't want your fence to collapse and your pigs to escape. It needs to withstand weather and any pig or other animal running into it. The good news is that they don't have thick fur, and they have wet noses, so electric fencing gets their attention.

We had our pigs set up in a forested area for brush-clearing, and a deer didn't see the fence in the middle of the night. The poor thing ran into it, then started to panic and jumped over it twice (once getting in, then she had to get out). She caught her hoof on the top the second time she jumped it. I was incredibly glad that I had researched the basic guidelines for secure fencing rather than just going out, buying some, and following the instructions. The fence held up. We had to tighten the fencing and stabilize the posts again, but they didn't fall, and no pigs got out. Making it stable and secure from the ground up is well worth the time and effort.

This footnote offers several electric fencing options as well as specific spacing instructions for containing pigs.11 Getting the height of the bottom two wires right is vital. Wires should be at about 10", 12" (or nose level), and 18." Go by the size of your pigs rather than measuring inches. You want them to get the "message" on their nose, and then be discouraged from going under the nose level wire.

Note that a pig will jump back when the electric shock is on their nose in front of their eyes, but they will lunge forward when the shock is behind their eyes. So, if they get their snout under the lowest hot wire and get shocked on their ears, their weight will lunge forward into the fence.12

Training Your Pigs to Respect the Electric Fence

You will need to train your pigs to respect the electric fence, and you want their first encounters with it to make a clear impression on them. The electric fence is not a physical barrier to the pig. It is a training device, so they understand that touching will be unpleasant and should be avoided. If you just put them in the pen with an electric fence, one of them is likely to run right through it, have a bit of an uncomfortable zap, but think, "if I just charge at it, I can get through to the neighbor's gardens – well worth a little zap." If your pigs learn that a slight touch causes pain, they will avoid it. However, if they breach it and figure out that running at speed will get them out quickly, then you will have a problem.

To train them, we tied some plastic ribbon markers along the fence to give them a visual cue so they would associate the markers with the pain. Then we supervised them and put some yummy treats inside the fence line. They could get the treats inside, but they learned that it hurt when they curiously sniffed at the fence. This practice formed an association in their clever pig brains that they did not want to touch that thing. After having that experience a few times, they did not think to try to go further. It was established immediately as a psychological barrier rather than a physical barrier that looks weak and easily breached (which it is).

Since I trained my own pigs, I came across this article by David Devold[13] regarding training pigs to respect an electric fence. He discussed similar methods that I was taught but adds an intermediary step. Before putting his new pigs out with the electric fence, he sets up a training pen with hog panels outside the electric fencing. The panels are stable, and if the pigs choose to try to breach the electric fence, they will never succeed. Contact with the fence never gets results, only an unpleasant "bite."

Getting the current level right is critical. I use a 15-joule charger. 15-joules is the standard recommendation amongst homestead pig keepers. Devold recommended getting a charger that will work for 2-3 times the amount of fence you think you will need.[14] His advice is wisdom from experience. You are most likely to end up wanting more fencing than you thought you would.

Having a mobile pig tractor is one of the most definite pluses of keeping pigs on a homestead. A "pig tractor" gives you the benefits of the pigs mowing and rototilling while being able to easily move them before they do damage. Learn about fencing, train your pigs, and ensure that they respect the fence. If they get out, you may not only lose your pig to a predator or a car accident; but they can also do a lot of damage to your gardens and your neighbor's property. In addition, if you do not catch them and they become feral, you may be responsible for contributing to the feral pig problem that destroys large swaths of landscapes.

What if...?

Another consideration for enclosures is the "what if." What if you need to isolate a pig from the herd? You'll need extra space for any new pigs to go through their quarantine time anyway. Even if you do not have this set up when you purchase your first pigs, think about how you would provide extra isolation space if you needed it.

Most pig owners who have mobile grazing areas around their land have a permanent barnyard enclosure near their house (but not too near because of the odor). This gives the option to watch over a farrowing sow, isolate a pig who is ill, or keep an eye on a pig in quarantine. Before they are let into the pen with the herd, you will need to temporarily fence them *next* to your herd so they can get acquainted. See the section on introducing a new pig for more details.

Supplies

Pigs do not need a lot of supplies, but there are a few.

Pig Water Troughs and Dispensers

Water is critical, and sadly, often overlooked. All mammals want clean, fresh water to stay hydrated, and pigs are no exception. The challenge with pigs is to provide a water trough that will not tip over, and keeping that water in the trough as pigs are... messy.

If you like a DIY option, the farmer who assembled this automatic waterer gives excellent instructions.

Sheraton Park Farms Pastured Pork Automatic Pig Waterer[15]

He talks about using a forklift to move it when he moves the pigs, but you could disassemble and reassemble it if you don't have a forklift.

Here are some other instructional videos for setting up mobile [16]pig waterers. This one is listed as "the easiest pig waterer ever!" It has nipples (a big plus); he gives good instructions, and I think I'm sold.

YouTube Channel "A Watts Way Farm" - "The Easiest Pig Waterer, Ever!"[17]

If it is possible for you to run a water line and use an automatic filler/dispenser, that is the best scenario. We've had both situations. For example, in our meadow, we could run a hose and set up automatic water filling, but in one of the forested areas, we had to bring water down in the truck and fill the water trough every day (or twice a day when the temperature got high in the summer).

An automatic filler with "nipple" dispensers is the deluxe version for a pig watering system. Not only is the water automatically filled; it is always clean when they drink it. You just need to ensure that each pig has a nipple.

Sometimes pigs who have not seen a nipple water dispenser might not know where or how to find their water and end up getting dangerously dehydrated. Observe them carefully. One recommendation is to wedge a kernel of corn in the dripper and let it drip into a pan of water so they can learn. [18]

Below is an image of the trough I used. It has the option of putting in a "float" system to set up the automatic filling, but it was easy to move without it. [19] You can see the optional float kit for automatic filling on that page as well. When I did not have the means to use the float and autofill it, I found it was stable enough so that the pigs couldn't knock it over. If I had more than three pigs, I would have invested in two of these to limit the spillage.

Rubbermaid Commercial Products Structural Foam Stock Tank[20]

If you don't want to use automatic waterers with nipple drips, then a Rubbermaid stock trough like the one above or a children's kiddie pool will work fine. You must change the water at least once, often more times a day, depending on how much mess they make in it. Installing a drain plug can be convenient. The trough above has a drain plug option.

Pig Feeders

Feeding homestead heritage pigs is different from farming for market meat production. You don't need to feed your pigs expensive grain. Therefore, we are not discussing pig feeders based on grain-fed market animals. Instead, homestead pigs can live on compost, vegetable peels, scraps, and the pasture/brush in an enclosure. We go into more detail on this in the next section. While some feed will be necessary, we are focusing on feeders for vegetables and compost when discussing feeders.

Many heritage pig keepers argue that you do not need containers for pig food at all – the pigs gobble their food up from off the ground. In fact, when you have excited pigs pushing each other to get to the food, it can be advantageous to pour it out in a pile for each pig along the fence line. They will often use their snouts to push the food up and out of the feeder onto the ground, anyway. (That's if the container doesn't get tipped over when they step in it.)

If you want to use containers, I recommend small feed pans like this.[21]

Little Giant Heavy Duty Rubber Tub Durable Rubber Feed Pan[22]

Before I decided to just dump the food on the ground, I used the one shown above. A big plus is that this feed pan will not degrade from freeze/thaw cycles. If you have only 3 pigs, get 3 pans for less drama and more chance that the container will not get tipped upside down. See the training section for making your pig feeding more accessible for the pigs, easier for you, and less dramatic, generally.

If your pigs are kept in a permanent enclosure, you can attach feeders to the fence.

TOP SELLER

TOP SELLER

Hulk Feeders for Swine
by Sullivan Supply

Hog Feeder w/ Door
by Little Giant

ValleyVet.com[23]

Hoof Trimmers

Pigs can become lame if their owners do not attend to their hooves. We will discuss how to do this in a later section. The grinder (file) is needed because the hooves are hard, and you can't do the job with trimmers alone.

My favorite pig hoof trimmers are shears. Here are two examples:

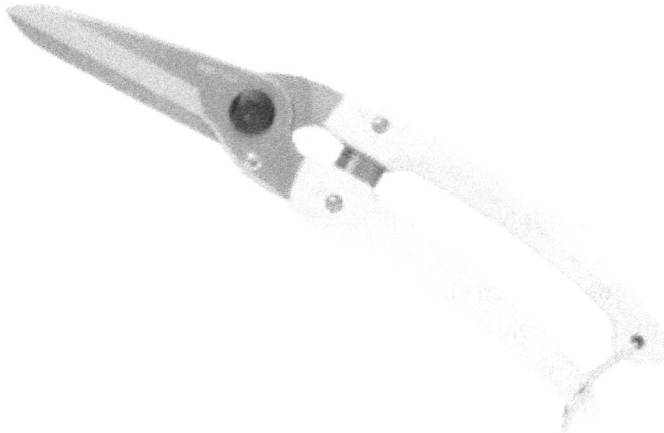

Premier1supplies.com[24]

A standard pig hoof trimmer amongst farmers are like these:

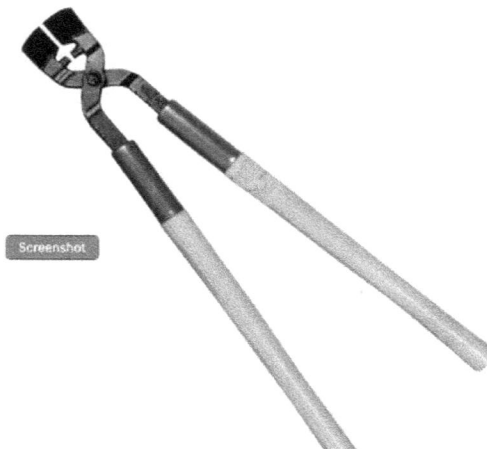

QCSupply.com - Heavy Duty Hoof Trimmer[26]

The heavy-duty hoof trimmers shown above are the same trimmers that my mentor used. I *could* use them, but I found that the shears were easier to use for my smaller hands and smaller pigs. If you decide to go with one of the larger heritage pig breeds, you may want to have the larger trimming tool.

Some farmers improvise with tools (you will see an example of this in a video demonstrating hoof trimming). Personally, I want my trimming tools to be sharp, so I don't have to struggle and apply a lot of force. Of course, you don't want them to be so sharp that they will be a danger to you if the pig moves its foot, but having dull tools will only stress you and the pig and make the job a lot harder.

Tusk Saw

For tusks, many pig keepers use a Gigli Wire Saw shown below.

Gigli Wire Saw[27]

Sawing next to a pig's mouth may require restraint that causes a lot of distress in the pig. There are other options. See the tusk section for this discussion, but consider the above veterinarian grade Gigli Wire Saw if you choose to have this as part of your pig toolkit.

Barnyard Clothing, Shoes, & Gloves

If you go inside the pigpen, you are likely to rub up against a pig. So, I kept a super large pair of pants handy outside that could go over anything I was wearing. I kept an old shirt as well.

You'll want knee-high rain boots, muck boots, or steel-toed shoes. I was wearing my Hunter knee-high boots until my toes got stepped on by a couple of excited 175 lb. pigs. They weren't acting aggressively, but they were excited and not trained. After that, I wore my steel-toed boots or shoes. I only used the rain boots when I needed them for serious mud.

Whatever shoes you choose, designate them as outdoors/barnyard only to prevent not only mess and smell in your house but also bacterial and parasitic contamination.

I like to use these heavy-duty rubber gloves for pig purposes.

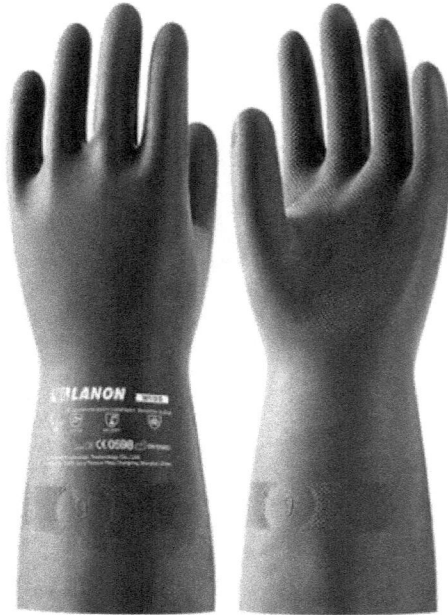

28

They provide protection while allowing for dexterity. In addition, leather gloves will soak in fluid from anything wet or damp (eww). I have a cotton-lined pair for summer and another pair that is fleece-lined for the winter.

Part Three: Maintenance

Diet for a Small Heritage Pig

This book assumes that you are interested in pigs for a homestead and will use them on your land as mowers, brush clearers, and rototillers. Therefore, we will focus on the homestead foods and conclude the section with grain supplement if you find it necessary.

Water

Water can be overlooked and underrated. Pigs require a constant supply of fresh, clean water. *Each* pig needs to have access to 5 gallons per day. If you have lactating sows, then they will need 6-7 gallons a day.[29] If you are using an open trough, then at the very least, water must be changed daily. If you are using a small trough and they step in it and get the water dirty, then you have to be prepared to fill it as needed.

The recommended best practice is to utilize automatic waterers, where the pig learns to turn the water off and on when they want a drink. Automatic waterers can use a "float" that indicates when the water needs filling or a "nipple" that is screwed onto the end of a hose or pipe. Do not use buckets for watering pigs, as their curious and playful nature will result in spillage and waste water. The importance of water for animals should be emphasized for your herd. Utilize these resources to have a better understanding of water recommendations for swine: [30].

If pigs do not get enough water and become dehydrated, they can suffer or die from salt poisoning. (Salt poisoning can also occur if there is too much salt in their feed.) Salt poisoning affects the central nervous system and often results in "meningitis-like symptoms."[31] You can find more details about salt poisoning symptoms here.[32] In addition to listing more symptoms, the author stresses that if you expect salt poisoning, do not allow the pig to have unrestricted access to water, as this is likely to make the problem *worse*. Slow and measured rehydration is critical, and your vet needs to be consulted.

As with most ailments, prevention is the best cure.

How much will my pigs need to eat?

Before diving into the specific foods for your pigs, let's discuss the amounts of food.
These numbers are based on the American Guinea Hogs - the Kunekunes need a little less, but these amounts will give you an idea.

- Adults need about 4% of their body weight a day. For example, a 120-lb. pig will need just under 5-lbs. a day (4.8 to be exact).

This number can get confusing when you have pigs of different sizes and ages. What about the young adults who weigh a lot less? What happens if one pig is pushed out of the feed?

There are a few things you can do. Some pig farmers set up a way to separate their pigs into groups before feeding time to deal with this. Our pigs had a lot of drama (and apparent food scarcity fear) at feeding time, so we decided to experiment with calculating how much food everybody needed and putting it in a line along the ground so that they weren't piling up on each other for one spot. That seemed to work out perfectly well.

The method described above might work for you, or it might not work in your herd. Watch your pig's weight (is anybody losing weight? Is anybody getting obese?), watch for bullies keeping one pig out of the food. When we were feeding over the fence line, we had one pig routinely pushed away, but it solved the problem when we spread the food out. Piglets seemed to grow and thrive just fine when the food was spread out.

If you notice bullying of a pig or piglets so they can't access enough food, you'll have to find a way to isolate the bullies away from them so they can eat. Some pig farmers have a system of rotation where they let one or more pigs into a feeding area, put them back, then let others inside to eat. Get local advice if you are having trouble with your pigs being overly pushy with each other at feeding time.

The Diet

Both American Guinea Hogs and Kunekune pigs need to eat lots of grass and pasture. In addition, they have different nutritional needs from other pigs, so do not follow general pig feeding guidance for these breeds.

The following bullet points list how we feed our American Guinea Hogs through the seasons. This is a common approach among owners of American Guinea Hogs and Kunekunes.

- Pasture is the main bulk of food *except for* November-March when it is not available in many climates. To determine the proportion of other foods required for your pig's nutrition, assess your own ecosystem for the seasonal availability of pasture. Alfalfa hay will constantly be supplementing the pasture to ensure proper nutrition. At least a little pig feed will also fill in the nutritional gaps.
  - This article describes a small pig farm in a northern climate that relies on pasture for the main feed. They even reached 100% pasture at one point. [33]
  - It is important to note that "pasture" does not just mean "grass." Grass is *one* of the components of pasture. Other plants might be legumes (e.g., alfalfa), clovers, millets, oats, barley, amaranth, chicory, burdock, and thistles.
  - One of the big reasons people supplement pasture with pig feed is to ensure that the pigs are getting the specific amino acids, especially lysine. Unfortunately, pasture alone is low in lysine. These pig keepers describe how they ensured the lysine content without pig feed. [34]
- Alfalfa hay
- Forest woodlot
- Pig feed or pellets[35]
- Vegetable scrap

Raw Vegetable Scraps

Most small homesteads need to supplement the grass. The use of raw vegetable scraps is widely agreed on in the pig keeper community. Many pig keepers get raw vegetable scraps from their neighbors, kitchens such as school canteens, rest homes, or restaurants. If you do receive vegetable scraps, make sure that you provide a clear list of things that cannot be included. During the spring and summer, the vegetable scraps from 8 people supplemented the pasture grass and rototilling. I deliberately made sure that I wasn't keeping too many pigs over the winters. I kept it down to 2-3. In addition, we arranged to pick up large amounts of scrap vegetables from a local store during the winter to supplement our own vegetable scraps.

Another source of food can be found in orchards in the autumn. Pigs will help you clean up the fallen fruit or nuts. There is a warning with apples and peaches that the pits contain cyanide. [36] If you have an apple or peach orchard and would like to let your pigs help with the clean-up, then talk to your livestock vet about how many pigs will be clearing how many trees. They may provide customized guidelines for rotating them in and out of the orchard a certain number of days, then giving a break, or including other foods to offset the toxicity of the pits from the stone fruits.

As you may expect, piglets have different needs for growing than mature pigs do for healthy weight and muscle building; pregnancy and lactating also make a difference. Let's discuss nutrition through the life stages of your pigs.

Piglets

The early growth to maturity is critical for a pig as it is for any mammal. Healthy development of the bones, muscles, nervous system, and organs all depend on it.

For this reason, some pig farmers recommend depending on a formulated pellet diet that ensures that the right nutrition is supplied to the piglets.[37] Others feel more confident easing their piglets into alfalfa hay and vegetable scraps while supplementing with pellets. The reason pellets are especially recommended at this stage is that they are more easily digestible and help their digestive tracts come to maturity after weaning.

I liked the approach of starting with the pellet for about a week, then starting to mix in fresh vegetable scrap, see how they respond, add in some alfalfa hay, etc. Of course, if my pigs get diarrhea, then I pull back the proportions. Thus far, I have had smooth transitions to wean them from a pellet-based diet.

Young, Growing Pigs

If the transition from pellets to other foods is gradual but steady, then the young, growing pigs will need to be monitored for protein content. Fortunately, the AGHs and Kunekunes require less protein than the larger breeds and certainly less than the commercial breeds.

High protein and lysine are essential for young pigs to grow.

Young Adult Pigs

If you want to get very specific and technical with nutritional needs, you can adjust the specific amino acids and lower the protein content as the pig grows older.[38]

Adult Pigs

Pigs that are over 200 lbs. should have less protein and lysine. [39]

Finishing

It is advised that the best meat will be produced if you back off on the protein intensity a few weeks prior to slaughter. This is the "finishing" stage. You can either get pig food specially formulated for finishing, or you can put them out on a wooded lot during that phase.[40]

Pigs have a reputation of being super hardy and able to eat anything. That's *almost* true. There is a surprising amount of conflicting information among pig experts about what pigs can and cannot eat, especially regarding food scraps. This can be very confusing. We will discuss the varied opinions so that you can do your own research, ask your vet's opinion, and make your own conclusions to decide what to feed your pigs.

What Pigs Should **Not** Eat

Note that there is a difference between something *fatal* (e.g., death cap mushrooms), *toxic* (e.g., green potatoes), and *ill-advised* for a variety of reasons, including meat taste effects and possible passing of pathogens to humans (e.g., meat, fish, milk).

Here is a list of foods that are often listed as being ill-advised for pigs. Few items will result in acute poisoning, but some can result in harm to humans. Also, it is worth noting the foods that can have slower, long-term effects on your pig's health. Where noted, there is conflicting information.

- Meat (this includes dog food)
  - Diseases such as foot-in-mouth disease. Several swine outbreaks were documented in the 1980s and traced to animal products in livestock feed. [41] It is noted that if pigs are fed leftovers that have had meat on the plate, anything that contacted the meat can transmit disease. [42] African Swine Fever is also a concern. Pigs can pass this disease on to other livestock as well as being infected themselves.[43]
- Fish (this includes cat food)
  - Even small amounts of fish scraps can make your pork taste fishy.
- Milk
  - Some experts list milk and whey as everyday things that people give their pigs and even report better growth and digestion when fed dairy in the winter.[44]
  - Other experts list milk in the "do not feed to pigs" list.

- o We included milk past date or beginning to sour for our pigs and had no problems, but you need to look at the evidence and decide for yourself. The laws for commercial pork not being fed milk products are related to mass production challenges.
- Raw eggs (cooked are OK)
  - o The concern is passing salmonella and other contractible diseases to humans.
- Sweet and salty foods
- Significant amounts of processed foods[45]
- Bread in large quantities as a bulk feed
  - o Your pigs will not get the nutrition they need from large amounts of day-old bread. This is not advised as a strategy.

- Walnut Shells
  - Some farmers say that they let their pigs out to eat fallen walnuts in orchards.[46]
  - Other sources say they are dangerous because the shards of the shells they crack can damage their pharynx.
- Tomato, cabbage, turnip, broccoli roots, and leaves
  - Lesser amounts of these vegetables are ok, but your pigs should not rototill a garden where rows of these were planted. Cooked turnips, for example, are fine to feed your pigs, but not the greens or the deeper roots.
- Pitted fruit, especially apples and peaches
  - You'll find pig keepers talking about letting their pigs out in their fruit orchards in the autumn and then other information pointing out that pits include elements that are toxic and dangerous to pigs.
- Green and Raw Potatoes are toxic to pigs
- Death cap mushrooms are fatal to pigs

Wild Plants to Avoid

Pigs can be wonderful partners in brush management. It is important to know what plants are toxic or poisonous to them. Some toxicity will be obvious quickly, while others are not detectable for a long time then suddenly fatal or a cause of serious illness.

- Hemlock
  - Like humans, small doses can be fatal. (In piglets, this can be within hours.) It can also cause congenital deformities.
- Bracken
  - Can cause acute heart failure and lung edema.
- Foxglove
- Henbane
- Ivy
- Cocklebur
- Laburnum
- Mustard seeds and roots
  - This is typically not found in woodlots, but don't choose pigs to rototill a meadow full of wild mustards

According to Dr. Georgina Crawford, the senior policy advisor for the National Animal Disease Information Service, the most common poisonings they are aware of are hemlock and bracken. [47]

Talk to your vet and homestead pig raisers who use pasture and woodlot about plants poisonous to pigs in your area. They will be a source of specific information for your ecosystem and save you a lot of time researching articles that include irrelevant information.

Table Scraps, Anaerobic or Moldy Vegetable Scraps – Yes or No?

For the most part, pig keepers agree that your vegetable table scraps from the day are OK to give your pigs as long as they do **NOT** have the following:

- No meat or fish
- No vegetables that have touched meat or fish (pathogens can be transferred)
- No highly processed foods/junk foods
- No foods that are high in salt or sugar
- No mold

Legally, the list of what you "cannot" have in your table scraps applies to the selling of meat, but not to private homesteads feeding your family. Having said that, plenty of other pig keepers advocate including table scraps for your pigs. People have lived with pigs feeding table scraps to their pigs, and in fact, still do. Taiwan is an example. Being a country with a small land base, they have been challenged to decrease their landfill requirements. They also recognized that the wet food scraps decreased the efficiency of their incinerators. Pigs happen to be Taiwan's most significant source of meat, so they have had a highly successful program encouraging people to feed table scraps to pigs.[48]

Penn State Extension suggests that table scraps should be fed to pigs only with "careful discretion." They point out that pigs have nutritional needs that would not be met depending on a lot of table scraps for their diet and that you run the risk of spreading Trichinella or African Swine Fever.[49] Then you have the people who advocate for table scraps and have fed them to their pigs without incident. [50] The UK even has a big project promoting the use of pigs to decrease landfill, "The Pig Project."[51]

Here is a vital resource published by the Harvard Food Law and Policy Clinic, Arkansas University, and the Food Recovery Project. [52] It's *called Leftovers for Livestock: A Legal Guide for Using Food Scraps as Animal Feed.* Even if you are not selling your meat, you can read the information and decide whether your small herd of homestead pigs is at risk for the particular concerns discussed.

Mold and Anaerobic (rotten) Food Scraps

Many pig experts will say that it is ill-advised to feed your pigs any rotten (anaerobic) or moldy vegetables. Here are the main reasons:

1) Disease-causing bacteria that can harm both pigs and humans (including mycotoxin poisoning)
2) Trichinella or African Swine Fever

All pigs, but especially pregnant sows and young pigs, are susceptible to mycotoxin poisoning. Mycotoxins are microscopic fungi found in molds. The damaging effects can sometimes be noticed quickly, but other times there will be long-term, chronic effects that are more difficult to diagnose.[53]
The common signs of mycotoxin poisoning are:

- appetite loss
- weight loss
- respiratory issues
- increased susceptibility to infectious diseases (poor immune function)
- poor growth rate[54]

When I was learning, I decided to pursue a course of common-sense approach. My pigs were getting high-quality food from pasture/alfalfa hay and pig feed supplemented with vegetable scraps. Sometimes they got a moldy strawberry or two, but I wouldn't pour a whole bucket of moldy food in their pen as their main meal.

I appreciate this advice from pig keepers: "don't feed your pigs anything you wouldn't eat."[55] This seems like a wise guideline. Still, there are plenty of pig keepers out there who will include an occasional piece of moldy bread or spoiled vegetables from the fridge when they feed their pigs.

Using the Manure

Manure is more or less of a problem depending on the size of your homestead and the number of your pigs in it. Manure management is one reason to start small with your herd.

Pig Manure vs. Other Livestock Manure

This section will discuss the management of manure on
- Pasture or Forest
- Enclosures with "feeder pigs."

Then we will cover composting methods and the level of risk for contamination when used as a soil amendment in vegetable gardens.

Pasture or Forest Fed Pigs

If your pigs are set out on a large enough pasture or forested area, you may be able to let the manure go into the land. This is the simplest, most elegant solution and an important consideration when deciding on how many pigs you have on one piece of land at a time.

Pigs will want to poop away from their food, water, and bed. Therefore, you will probably find a dense area of poop in one place that needs to be spread out.

You must think about what you want that pasture to be when the pigs are rotated to another place. Pig manure is lower in nitrogen but higher in acidity than other manures; therefore, it can be beneficial or a hindrance to a pasture space, depending on your intention.

In a large pasture with a few pigs, I observe their habits, then spread the manure out over the whole pasture when I rotate them to another area.

Feeder Pigs Inside Enclosures

This is the most difficult as you must deal with **all** of the manure, all the time. You have two choices:

- Use a deep litter system
- Clean out the enclosure as needed to ensure that they are not forced to live too closely in their poop. (This may be daily, depending on the size of the enclosure and number of pigs.)

A deep litter system is when you keep putting down thick layers of fresh dry straw on the bottom of the pen. The manure starts to compost (which is good), and you don't have to shovel every day.

- The downside of a deep litter system is that you have a *lot* of heavy, wet manure that is partially composted to deal with every few months. However, a lot of people who pasture their pigs use barnyards with a deep litter system in the winter.
    - Before you decide whether to do this, consider how you will move the compost and where you will put it. If you have a place to start a big compost pile, it's great. Do you have a tractor to lift it? Can you hire a piece of equipment or hire a local person to do it? If you are shoveling, it requires a lot of back strength.

Cleaning out the enclosure daily or as needed may also require equipment. Once, we had four pigs who were finished with their rototilling purposes on the land. To avoid damage to the pastures and woodlot they had grazed on, we put them in a large barnyard enclosure for finishing. The barnyard was about an acre and had enough space to allow them to make their latrine and not have to clean up after them every day. We spread out the manure overlaying with straw every couple of weeks to limit the latrine area's damage (and smell). They spent a couple of months there, and the trees growing there were fine after the pigs were gone.

Composting Pig Manure

There are varied opinions about whether pig manure is safe to add to a vegetable garden after composting since diseases, pathogens, and parasites that are harmful to humans and can infect vegetables through the soil. For this reason, some pig keepers say that you should not *ever* use pig manure on vegetable or fruit gardens, *even after it has been composted.* Their advice is to use it as a soil amendment for landscaping only. Others say that pig manure is safe (with only a slight possibility of something harmful being alive) after composting that is long enough and hot enough. The rationale for this is that the compost must be long enough and "hot" enough to kill all harmful components. Even after composting, it is claimed that some microorganisms have been known to survive, get into the soil and contaminate the food.

The Northern Nevada Horticulture Organization (NNH) promotes science-based gardening education. Here are some points from an excellent article on this subject:

- In 2002 researchers suggested that un-composted pig manure at 77 degrees F for 90 days would be free from pathogens. *Even so*, they added that it is recommended that you take caution regarding root crops and leafy vegetables (like lettuce where the edible part touches the soil) because there was still a risk of health problems.[56]
- Ron Becker of Ohio State University asserts that raw pig manure is safe as long as it is worked into the soil for at least 6 months prior to planting. This safety includes the "*...edible portion that is in direct contact with the soil.*"[57]
- NNH suggests following the American Organic Standards and Guidelines for Composted Manure. A temperature of 131-degrees F must be maintained for 15 days. This is deemed safe as it will kill E. coli Cryptosporidium, Salmonella, etc. Note that while this temperature is what it takes to be safe, manure compost will not be filled with the beneficial microbes that vegetable scrap composting can have. It's not too hard to maintain this for 15 days. You can put a thermometer in it and check it to make sure it is staying constant for just over 2 weeks in the summer, and then you're done.

One pig keeper told me that he knew a farmer in France who said that the French opted for time-based rather than heat-based for the biosecurity of their compost. So they will make a compost pile without taking its temperature for 2 or 3 years. So once you get started, you have a renewable source of safe compost every year.

- I do love using all parts of the life cycle of animals as a circular, sustainable existence.
- I do not love the risk of serious health problems.
- I do not love trying to maintain a compost pile with a heater.

My solution to this dilemma has been to make a livestock manure pile that is 1 part manure to 2 parts straw or dried leaves, then cover it with plastic sheeting (the kind used on greenhouses). In the summer, it gets hot, and there will be at least one period of 15 days straight that the compost under the greenhouse plastic will heat up to 131-degrees F and above. The downside of this is that depending on your climate, you may need a summer to get this kind of heat reliably. If so, it means that you cover it in the late autumn, go through an entire year and then plant the following spring.

If you want a more immediate option, then use a thermometer and set up outdoor heating lamps to keep the temperature up.

When you are making your compost pile, layer the manure with the straw and leaves. When it is time to cover it, turn the compost over to aerate it. This can be done with a pitchfork - lifting and turning chunks over to get some air circulated throughout. After you cover it in the late autumn, you can leave it for the winter if you are in a cold climate. If you live in a warm climate, you will need to do more turning to ensure that it has enough oxygen for the process to become healthy compost.

Buying your first Pigs

When you've decided which breed you want, we highly recommend going to the association for the breed you are choosing to find a reputable breeder. When you contact the breeder, ask whether their pigs have the official papers for the pigs. As well as ensuring that you are getting what you pay for, these heritage breeds need tending and best breeding practices to ensure the genetic quality and health of the breed in the future.

There are some qualities to look for in all pigs, but some breed-specific traits as well. So first, let's look at the characteristics for breed standard AGHs and Kunekune pigs; then, we will turn our attention to a checklist of things to look for in any heritage breed.

Another aspect to look for in the pigs you are purchasing is whether they have been handled and whether their hooves have been trimmed. Observe the pigs you are considering purchasing; how do they respond to the farmer? Can he walk up to them and pet them? How do they respond to you? Ask the breeder directly whether the pigs allow their hooves to be touched and trimmed. If you are purchasing a pig whose hooves need trimming and they are not used to it, you need to know that.

The compact, genial Guinea hog may bring your family
pleasure as well as pork.
Photo by Lynn Stone

Mother Earth News Blog
"American Guinea Hog: A Small Pig Breed for Homesteaders"[58]

Ears on an AGH stand up. A healthy adult **may** have ears that fold a little bit forward at the very tips, but not much, and not always. If the ears are not upright, don't purchase the pig.

Their snouts are medium-sized. If the snout is short, like a Kunekune, or elongated, this is not a good sign.

Most AGHs are solid black. It is OK if there is a little bit of white on the nose or feet, but not much. There is a rare gene that sometimes emerges that results in red color. Ask specific questions about the genetic line of the pig in question.

Kunekune Pigs

The US American Kunekune Pig Society adopted British standards to establish the US breed standards in 2013. [60]

The snouts are short, but the shape of the face should not interfere with their eyes.

Ears may be upright, tilting slightly forward, or be floppy.

Jowls should be small/medium. Huge jowls would indicate obesity or a genetic trait you do not want to breed.

The color should be one of those listed on the AKKPA color chart. [61] These include:
Cream, Ginger, Black, Brown, Ginger/Black, Black/Ginger, Tricolor, Black/White, White/Black, Brown/White, White/Brown, Gold Tip

If you choose a heritage breed other than the AGH or Kunekune, check the breed standards to know what to look for.

In addition to traits specific to the breed, here are things to look for in any heritage pig:

- Check to see that the shoulders are even
- The chest should be wide enough to fit between the legs
- Buttocks should be well rounded

- Both females and males should have 10 teats. (Yes, males have teats.) They should be spaced evenly and look healthy.
- Watch how they move. Can they run comfortably? Find out what they have been eating. If they have been fed grain, then it is easy to overfeed them. Obesity in AGHs can cause infertility.
- Get official papers and health documents.
- Look for any health-related signs like discharge from the nose or eyes.
- Ask about vaccinations and de-worming that has (or has not) been done.

Maintenance

Pigs are relatively hardy and easy to care for compared to other livestock, but there are important considerations for their health, safety, well-being, and the pig keeper's sanity. We will discuss the following:
- Health and Care
- Pigs Need a Job
- Predators
- Training

Health and Care

Prior to getting your pigs, find a vet who has swine experience. A knowledgeable and experienced vet will give you guidance for the appropriate vaccinations and medications necessary for treating and preventing illness.

If you do not have access to a vet that has experience treating pigs, contact the Penn State Extension Educators. [62] This site encourages pig keepers to contact them, especially if you cannot find a qualified vet nearby.

Vaccines

Purchase your pigs through a reputable breeder as recommended. Then, you should have the quality assurance that the pigs were from **one** farm only (not purchased elsewhere to be sold again), and if the pigs are past 8 weeks old, there should be vaccination papers from a vet.

Below you will find the vaccines recommended or to be considered for pigs. These are basic and are often administered in combo-vaccines. There are regional variations and risks that may require supplemental vaccines, so vet consultation is critical to keep your pigs healthy. [63]

- **Erysipelas (aka diamond skin disease)**
  - Erysipelas can be fatal. Symptoms are a high fever and bruises on the skin that take the form of a diamond pattern. The bruises come from the blood vessels being compromised, which causes internal bleeding. It is also very painful.
  - Your pigs will definitely be exposed to erysipelas as birds carry it, and it lives a long time in soil.
  - Vaccinate your pigs Erysipelas at 8 weeks, then give them a booster in 2 weeks. After that, once a year.

- **Tetanus**
  - Tetanus also lives in the soil, so pigs are susceptible to this bacterium as well just because they root and graze. They *tend* to have a resistance to tetanus, but some contract it.
  - If pigs get tetanus, treatment is usually not effective, and it is fatal.
  - Vaccinate your pigs at 8 weeks, then get a booster in 2 weeks. After that, every 6 months.

- **Leptospirosis**
  - Leptospirosis is a bacterium that is transmitted by wildlife, usually through waterways and even puddles. Mice urine can carry it; raccoons are a sure sign that you have leptospirosis on your property.
  - It is contagious to humans.
  - The pig contracts the bacteria, and it becomes a permanent carrier. The microbes infect the kidneys; this is why it is carried through urine. Your infected pig will infect your whole herd.
  - Symptoms in pigs include infertility, abortions, or stillborn piglets. Other health issues can exhibit these symptoms. These symptoms can be misdiagnosed for something else; always get professional advice from your veterinarian.[64]
  - Symptoms in humans can include the following symptoms, but according to the Center for Disease Control, some people have no symptoms at all but are still carriers. [65]

    - High fever
    - Headache
    - Chills
    - Muscle aches
    - Vomiting
    - Jaundice (yellow skin and eyes)

- Red eyes
- Abdominal pain
- Diarrhea
- Rash

o You can see how these symptoms can be mistaken for other things. This is why it is critical to ensure handwashing after handling pigs or being around water sources. The disease can spread from the barnyard to the whole family.

o Vaccinate your pigs at 8 weeks, then get a booster in 2 weeks. After that, every year.

- **Actinobacillus Pleuronpneumoniae (APP)**
  o Commercial pigs confined indoors are susceptible to this disease; however, it is recommended that all pigs used for breeding be vaccinated. The piglets will have immunity, and you can prevent outbreaks in your herd.
  o APP causes lung damage and is often fatal and often exhibits as sudden death with hemorrhaging from the nose. If the pig survives, they are likely to have a compromised respiratory system.
  o Combination vaccines for pigs commonly include this.
  o Ask your vet about injectable antibiotics you might keep on hand in case you see symptoms.
  o Pigs 8-16 weeks are the most likely to contract APP.
  o The most likely cause of APP in homestead pigs will be the introduction of an infected pig or humans with it on their boots or other clothing. Dehydration can play a part; even "temporary water deprivation will trigger the disease." [66]
  o The two main ways to prevent APP are:
    - Ensure that any pigs you purchase have come from one farm that has good health history.
    - Quarantine new pigs for 6 weeks, watch for any symptoms, and during the quarantine, disinfect your gloves, boots, and any clothing that has come in contact with the pig or material in the pen before using them in your herd.
  o Symptoms include
    - Difficulty breathing
    - Blue tinge or patches on the ears
    - Lethargy
    - Lack of appetite

Rabies

- Rabies is different for pigs than for humans or pets. Commercial pigs are usually not vaccinated for rabies because they are not exposed to it in confinement. Homestead pigs, however, could contract rabies from wildlife or a domestic dog.
- If your pig bites someone, it must be reported to the health department. The test for rabies on pigs is fatal.
- Consult your vet about whether they recommend a rabies vaccine for your pigs. Since ours were sometimes in forested areas, we included rabies in the vaccine regime.
- Rabies vaccines are given to pigs at 4 months, then a booster in a year. After that, every 3 years.

Most veterinarians recommend a combination vaccine for homestead pigs with a booster schedule. Whether they recommend including rabies depends on the risk your pigs might be.

Common Diseases and Ailments

Even though pigs are generally considered "tough," pig keepers are wise to keep an eye out for these health conditions:

Review of previously discussed conditions:

- Dehydration, which leads to salt poisoning, is sadly, common and easily preventable danger.
- The following diseases (discussed above) can be prevented with vaccines and biosecurity precautions:
    - Erysipelas (aka diamond skin disease)
    - Tetanus
    - Leptospirosis
    - Actinobacillus Pleuropneumonia (APP)
    - Rabies

Additional health conditions to watch for:

Parasites

Pigs live a lot of their days with their snouts and mouths in the soil, finding food. When they aren't eating, they are rolling or lying around in it. They have (thankfully) evolved to handle many parasitic passengers, but sometimes even pigs get overloaded and succumb to ill health.

Here is a list of the most common ones that might infect a homestead pig. Several of them come specifically from feces, so if you keep your pig enclosure cleaned up, that will lower the risk.

Large roundworm (*Ascaris suum*)

Roundworms infect pigs through the feces of another infected pig. Purchasing tested or de-wormed pigs to start with a clean herd will prevent this. After that, when you introduce a new pig, talk to your vet about de-worming before the new pig's quarantine is up.

Roundworms can be very damaging or fatal to a pig. They attack the liver then move into the lungs. Piglets under 6 weeks old are the most vulnerable to roundworms as their potential resistance has not built up yet. Pigs with roundworms may die of severe liver disease or pneumonia.

Swine whipworm (*Trichuris suis*)

Whipworm is another parasite contracted through infected feces. It establishes itself in the intestine and causes diarrhea with blood and mucous. If you see this symptom, isolate the pig immediately and contact your vet for treatment. Your entire herd will need de-worming for this.

Lungworms (*Metostrongylus*)

Pigs ingest larvae through infected earthworms. The larvae hatch and make their way to the lungs and mature to adulthood, and hatch eggs. The pig then coughs up the eggs, which the earthworms eat, and the cycle continues.

An elegant but deadly circular life cycle. Lungworms can cause pneumonia in pigs.

Scabies (*Sarcoptes scabei* var. *Suis*)

Scabies is caused by a mite. It is highly contagious and is transmitted by an infected pig. The skin becomes red with some raised areas; there is hair loss and crustiness around the eyes and snout.

Prevent scabies with best practices when you purchase your pigs and through quarantine observation when you introduce a new pig to your herd.

Dippity Pig

Dippity Pig is pig shingles.[67] The symptoms are red inflammation on the back with lesions running across sideways on the back. There is no treatment, but fortunately, it only lasts about 24-72 hours. The pig is very susceptible to sunburn at this time, so keep them in deep shade or in the dark and let them rest with plenty of water. Offer food, but they may have a loss of appetite.

Urinary Tract Infections (UTIs)

Male pigs are most prone to UTIs. It is important to catch them as (like humans) if not attended to, they can spread to be a bladder or kidney infection. Like any mammal, if you see a pig trying to pee (often exhibited as "pee-stop-pee-stop-pee-stop..." then give them some pig cranberry tablets. I like to keep these on hand. I have had to treat UTIs twice. Both times they went away without becoming more serious.

Uterine Tumors

If you are not going to breed your female pigs, they need to be spayed. Un-spayed females tend to get uterine tumors that can grow very large and cause health issues. Talk to your vet about spaying females who are not going to breed and let them help you make this part of your cyclical planning through the year if you do decide to include breeding in your pig maintenance.

Obesity

This is all too common, and there's no excuse for it. Your pigs need to have a healthy diet in the right amounts for their needs.

Do your pigs eat any of the following?

- Corn will make pigs tend to gain a lot of weight
- Large amounts of left-over bread
- Processed foods
- Feeding too much

Obesity is preventable. It can cause infertility, similar organ problems that humans can have, and other conditions such as entropion which we will discuss next. Obese pigs are not better to eat. Like humans, pigs easily put weight on, and it takes a lot longer to get it off.

Pay attention.

Entropion

Entropion is a distressing condition characterized by the eyelashes turning in instead of out. They then scrape the eyeball. Sometimes this is genetic, but other times it is caused from obesity.

The only treatment for entropion is surgery. Do not breed this pig.

Conclusion – Common Diseases and Ailments

The list discussed above is not exhaustive. It is intended to give you a starting point so that you have some basic knowledge when you speak with your vet.

We encourage you not to be intimidated by this list for two reasons:

1) If this has been new and overwhelming information to you, consider that the lists of ailments for a dog or cat are surprisingly long. We are accustomed to primary care, prevention, and treatment for dogs and cats, so it doesn't feel as intimidating as it might for pigs.

2) Note that most of these things are preventable if you know about the risks and design your pig care protocols to support a healthy environment for your pigs.

- Know the signs and symptoms of diseases and conditions
- Have pig cranberry tablets handy, along with any other medications recommended by your veterinarian
- Be prepared to quarantine if necessary
- Contact your veterinarian for treatment and advice
- Prevention is always the best cure

Pig Spa Day: Hoof and Tusk Trimming

Hooves

Hooves need to be trimmed because they grow out and affect the pig's gait. When this happens, it affects their muscles and joints and eventually can lead to lameness. The hooves should be flat and even, without extra growth on the side or in the middle. Dewclaws, the extra hoof above the portion that comes into direct contact with the ground, should also be trimmed as they can grow to inhibit their walking.

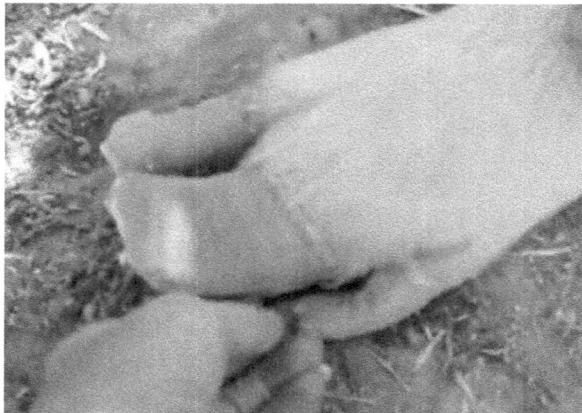

Overgrown pig hoof[68]

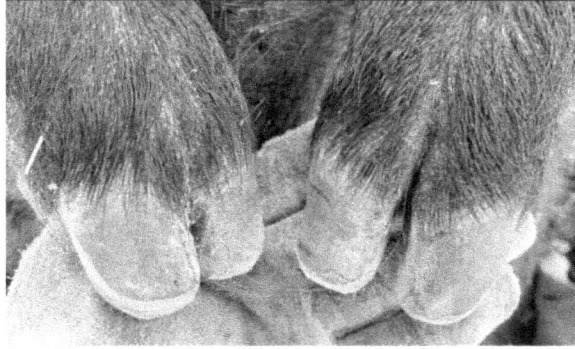

Healthy, Trimmed Pig Hooves

AmericanMiniPigAssociation.com[69]

Even if you are going to slaughter your pigs at 6-9 mos., you may need to do some trimming any time after 6 mos. This is one of the critical reasons that you should train and handle your pigs. They will relax and cooperate with trimming if they trust you and are used to being handled.

First, it is necessary to understand the anatomy of the hoof.

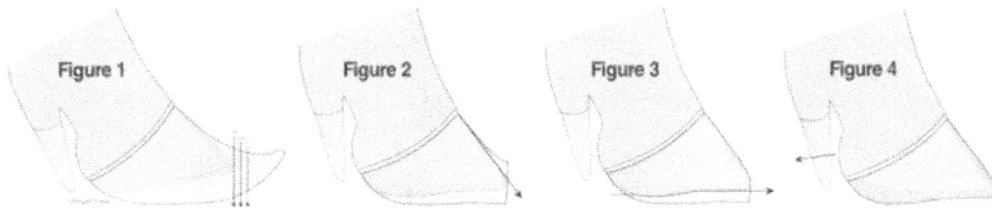

Understanding hoof anatomy is important for effective hoof care and treatment.

Illustration from Wattanet.com – Sow Hoof Trimming Guide[70]

The training section will go into more detail about how to train your pigs so that you are set up for "pig spa day" success. Let's look at the specifics of the hoof trim session:

This video from the Central Texas Pig Rescue is a wonderful demonstration of hoof trimming.[71]

YouTube Channel "Central Texas Pig Rescue"
Video is "Hoof Trimming"[72]

Notice he spends a little time with the pig and brushes him to relax. As the handler brushes the pig, the pig automatically lies down on his side and relaxes. The handler pets him, scratches behind his ears; the pig clearly loves this. The handler is calm and gentle.

Watch the way he trims not only the length of the hoof, but also the sides where there are bits that have grown unevenly. Also, one of the hooves was longer, so he ensured they are even.

Filing is necessary so that the hooves are truly flat and not set up to produce uneven growths.

This video is a demonstration with a pig whose hooves have grown far too long, as well as a pig who is used to getting their hooves trimmed. On the first pig whose hooves have not been tended, you can clearly see that by this time, it has affected not only the pig's gait but also the way she stands.

Still image from YouTube Channel "Teresa Johnson"
"Trimming a Pig's Claws... YES, you really can trim a pig's hooves."[73]

In the video discussed, the pig whose hooves need regular trimming literally puts her foot out for her handler when her belly is rubbed. *That* is excellent training and rapport! The pig who has overgrown hooves lies down on her stomach rather than her side and allows the handler to trim the front hoof right next to her mouth. She has never had her hooves trimmed and is occasionally falling asleep in the process. You want to aim for this level of trust.

The handlers in the second video underscore that stress is really hard on pigs. They say that if you must chase and "hog-tie" them, you are likely to have a sick pig in a few days. I will take their experience and skill as an authority on this. Pigs are extremely sensitive to stress. Watch the patience of the handler and the way he speaks to the pigs. If the pig will not cooperate and lie down, they advise leaving the trimming for another day.

Hooves that are overgrown require a larger tool and *much* more force to cut them. The handler is using all his strength and all the leverage he can get to clip them off. This is yet another reason to be observant and not let the hooves grow too long.

I was fortunate enough to know pig mentors who could walk me through this process. I trained my pigs per my mentor's instructions which were almost identical to what you see in the videos.

Two things the video did not mention, which I was taught:

1. Always carry the hoof trimming tools with you when you visit them for scratches and belly rubs.
2. Always handle their feet.

I gave them mini hoof massages, so they were used to the touch and had come to love it.

If you do not have access to a mentor who can walk you through hoof trimming the first time, then get advice from your vet and make sure you do the training to prepare the trust required to have a stress-free pig for "spa day."

Tusks

I never felt the need to trim the tusks of my pigs. According to the American Mini Pig Association, small heritage pigs like the American Guinea Hog and Kunekune will not need their tusks trimmed until they are 2-3 years old, *if ever*.[74] None of my pigs were around that long. Females' tusks stop growing when they reach adulthood, but the males will continue to grow.[75]

If you do see tusks that are either dangerous to people or other pigs or could get caught on fencing (*especially* electric fencing), then you may decide that it is prudent to trim your pig's tusks. In older boars, the tusks might get so long that they cause the cheek to bleed.

If you find it necessary to trim your pig's tusks, there are several options and opinions about the best course of action:

1) Use clippers or a garden tool.
- The clippers must be sharp so that it cuts the tusk rather than bending it which can result in cracking.
- This video demonstrates tusk clipping without restraint, stress, or a wire saw.[76] They are using a garden tool rather than clippers but are very conscious of not stressing or hurting the pig.

2) Use a Giglio Wire Saw

- This is the preferred method of some pig keepers. The American mini-pig association provides this information and video to demonstrate the use of a Giglio Wire Saw.[77]

3) Call the vet to do it (and show you how if you want to learn)

- There is nothing like a mentor for this procedure. Using online resources or books is a great way to take in some information so that you have some context when you get hands-on mentorship from an expert to take you through the process and watch you do it.

As with hooves, there is a "pulp" or "quick" inside the tusk. If cut, it will bleed. The pulp usually ends about an inch beyond the mouth, but individuals vary. The risks associated with cracking the tusk are significant because a crack in the tusk can cause an infection in the mouth. If you choose to trim tusks yourself, my strong recommendation is that you first watch videos of pig keepers explaining how to do it, then ask your vet to come up and show you how to do it. After that, you will be better prepared to do the job while avoiding dangerous consequences.

Introducing New Pigs

Now that you've read the health section of this book, you most likely already understand the importance of biosecurity on your homestead and the risks of infectious diseases and parasites being introduced to your herd.

As discussed previously, you need to have a place to isolate a pig(s) or a farrowing sow with piglets. This enclosure is also necessary for the quarantine of new pigs introduced to your homestead. Besides the biosecurity issue, there is a social issue as well. When your pig has gone through quarantine, you don't want just to put them in the enclosure and walk away. Let's go through the steps:

Quarantine

The safest recommendation for quarantine is 4-6 weeks to ensure that the new pig is not carrying something that could be transmitted to the herd. If possible, it is best if the pigs can see each other, but not make any direct contact.

During the quarantine period, watch for the symptoms of the health conditions discussed earlier. Always disinfect boots that have entered the quarantine area and any clothing that came into contact with the pig or objects (such as feeders and water dispensers) that it has touched.

When you've decided on a pig to purchase, ask the breeder what the pig has been eating. You want to do some transition feeding so that there isn't a sudden change in diet that causes diarrhea which could indicate a disease or parasite. Feed the same feed from the breeder along with a mix of the food you give to your herd. (Start with ¼ of your food, then ½, then ¾ till the transition is made.)

If the pig develops diarrhea, contact your vet. Many different factors can cause diarrhea, and you will need a professional diagnosis.

Remember to disinfect your boots, clothing, and don't use tools or supplies used in the quarantine area inside the enclosure with your herd.

After your pig's health is deemed healthy and it's ready to be introduced, put the pig in a pen next to the herd.

Pigs need to meet, get used to each other, and then integrate the new pig into the herd and re-establish the hierarchy. Let the pig live on the other side for a week or two while they get used to each other.

After a week or two, then let one into the pen with the herd. You need to know who the dominant sow is (or boar if you have one). In order for your new pig(s) to be accepted, the pig on the top of the hierarchy must signal a "yes" to the rest of the herd. Let the new pig in for about 15 minutes, then bring them back to their pen and continue this routine every day.

They may ignore each other at first, but don't be fooled. At some point, there may be some fighting as they establish the new hierarchy. I recommend having another person or two available to help you if they get into a fight that looks like somebody will get hurt. With Kunekunes and American Guinea Hogs, most of the time, this fighting will not result in any injury, and everything will settle down as the new pig is integrated. But be prepared to intervene if necessary. Separate them for the rest of the day, and then let the new pig back in again the next day.

My pigs did a bit of fighting, but it was more like posturing - then it was over. This pig keeper has great advice for working with another person: One of you herds the new pig in one direction, and the other person goes in the opposite direction with the pig from your herd.[78] The author points out that to do this, you (and your helper) must be confident and must already be established as "top pig" over the most dominant pig in the herd. See the training section for more details on this.

Contact your vet or an experienced pig mentor if you are having trouble. It is most likely that using the protocols described, your new pig will integrate into the herd safely and with relative ease.

Pigs Need a Job – Bored Pigs are Troublesome

The fact that pigs are multi-purpose creatures is a huge advantage on a homestead. They will provide meat for your family and can do a lot of jobs. A pig with a job is a happier pig, so it is a win-win.

We will be discussing the main functions of pigs on a homestead.

Pasture Mower

American Guinea Hogs and Kunekune Pigs love grass. Pasture mowing is the most natural thing in the world for them to do. If you only want an area to be "mowed" rather than also rototilled, keep a careful eye on when they are done with the top grasses and move them to another area.

Pigs will mow pasture beautifully, but they will also make a "wallow" (a depression in the soil to roll in mud or dirt). So after your pigs move out, you will have a dished-out area of compacted soil where they have been wallowing.

Depending on the time of year that your pigs are out on the pasture, their hooves may make depressions. If there is a lot of rain and the soil is soft, your pasture will be mowed, and it will come back, but there will also be holes. Ensure that the number of pigs you have to match the amount of pasture they are feeding on for the time of year and levels of rainfall.

Pigs will also make their latrine somewhere on the pasture. You will probably want to clean up this area when they are rotated off. On pasture that was not being used for human consumption, we just spread the poo all over the field.

If you put a mobile shelter in the pasture area, you will need to rotate it or be left with a big spot where the shelter's footprint was. You can also re-seed this small area with pasture seed to make it recover.

It is wonderful to have pasture grass maintenance.

Rototilling

This is the best job that our pigs did for us, and the story is an example of how pigs can do powerful work that other eaters (like goats) don't finish, and humans alone find impossible.

The previous owner of our land was elderly and had not tended the land in some time. As a result, the invasive noxious weeds were rampant in the meadows, especially what we now call "the pig meadow."

This meadow had knapweed that was literally 6' tall with flowers as big as my fists and stalks as thick as my wrist. I'm not exaggerating. The roots ran deep and thick. There were sections of the thicket that were so dense that it was impassable.

We put the pigs in the area for a while and they were working away at the weedy areas that were about a foot high, but ignored the tallest knapweed. So, we decided to chop down the biggest knapweeds and trim down any that were over 12." The trimmings were chopped into smaller pieces and fed to the goats.

Trimming the taller plants and chopping the stalks of the huge ones down to about 6" did the trick. The pigs first ate the green on top; then they started to root.

We allowed 3 pigs (and one litter of piglets till they were weaned) to be on that patch (about 1.5 acres) for two seasons. This is longer than what would be usually advisable, but we could see that they were still rooting down deep for food, so we let them go for it.

It became obvious when they were finished; they were only eating what we fed them. When we rotated the pigs off that patch, we had a messy, rototilled dirt area. There was not a speck of green on it, and it was very chopped up.

Knowing that the disturbed soil is an invitation for weeds to take hold again (especially knapweed), we seeded the area with native grasses and wildflowers. We scattered the seed after a rain; then, we rolled the area with the roller shown below. The rolling evened out the choppy, rototilled soil.

The grasses and wildflowers came up, and after two years, **not one knapweed plant** emerged within the pig enclosure. Instead, the pigs had eaten *all* of those deep roots that are so tough to eradicate.

This is our meadow in May before the wildflowers come up in the early summer.

This pig-powered knapweed eradication success contrasts with my efforts in another area of the land. There's a place with gargantuan knapweed that grows around our vegetable garden. For 4 years, I have dug with a shovel. The first year I dug them out as far as I could follow the roots. In the second year, I thought I had won as they didn't come back, but the third year they emerged again. I continue to work on it, it gets *better* every year, but my performance pales in comparison to the pigs.

It is critical to follow up with reseeding when you task your pigs with rototilling. If you don't, you'll end up with an invasive weed field. On the other hand, if you choose the right seeds for your ecosystem and intentions, you will find yourself with fertile soil for the plants you want. I opted for the roller, but another common practice amongst homesteaders is to seed the pasture and let the pigs stomp around on it for a day to get the seeds in.

Forest Maintenance

You may have seen "forest-fed" pork popping up in the meat section of your grocery store. Forest-fed pigs are increasingly common in the small-scale artisan meat industry as well as popular with homesteaders rotating their pigs through pasture and wooded areas.

Farmers in Europe have fed and used pigs as part of forest management for centuries.[79] The modern term for this practice is silvopasture; the old term was *pannage.* The critical component for silvopasture or pannage is that the pigs are allowed to forage in the wooded area for a *brief* period before they start rooting. If they start rooting, they will damage the root systems of the trees. Therefore, pigs rotated properly through woodland can enhance the health of the trees.[80]

Let's discuss the vocabulary terms specific to forest-fed pigs:

- **Silvopasture (in old Europe known as "pannage")**
The USDA defines silvopasture as *"The deliberate integration of trees and grazing livestock operations on the same land. These systems are intensively managed for both forest products and forage, providing both short- and long-term income sources."[81]*

As mentioned above, the trees can benefit from foraging pigs when managed correctly.

- **Masting**

"Mast" comes from the old English word "maest," which is food for swine. Masting in a forest is when the pigs are set out for short periods to clean up the brush and do very limited rooting so that the root systems are not damaged. Mast in a forest is the fallen logs, branches, scrub brush, acorns, pinecones, pine needles, leaves, etc.

Silvopasture is the modern, scientific, and technical term for masting or pannage that has been done for centuries.

- **Pastured or Grazed Woodland**
This is a term defining the areas where livestock have been kept. *"Forested pasture and range consisting mainly of forest, brush-grown pasture, arid woodlands, and other areas within forested areas that have grass or other forage growth."[82]*

A significant number of pig farmers and homesteaders are successfully sending their pigs out to forage in woodland areas and not damaging their trees. Fortunately, some of them share their experiences and provide videos for those interested in forest-fed pigs.

John Suscovich offered this excellent video. He demonstrated how he manages his pigs in a forested area, including how he moves his fences. He explained the whole project in the context of a holistic and long-term permaculture design view; "what do I want this to be in 10 years?" Asking yourself about your future plans is a great way to consider the role of your pigs in the long term on your land. This video is well worth the 4½ minutes to watch.

YouTube Channel "John Suscovich"
"How to Raise Pigs in the Woods"[83]

In the following video, Diego Footer interviews Luke Groce, who offered a detailed explanation with drone footage of where the pigs have been and where they are going. Then, he rotated through both pasture and woodland.

YouTube Channel "Diego Footer"
"Free-Range Pigs-Raising Pigs in the Woods and On Pasture"[84]
Interview with Luke Groce

Take a look at the following interview by Josh Sattin with the farmer, Bobby Tucker, who uses his pigs in a dense area where he is planning ahead to run his sheep. "Forest Pigs with a Permaculture

Expert."[85]   This is an excellent example of holistic permaculture planning and use of animals on the land as your partners.

YouTube Channel "Josh Stattin"
"Forest Pigs with a Permaculture Expert"[86]
Interview with Bobby Tucker

You will need to supplement your pigs (as always) with some feed for their nutrition, but you'll hear the farmers in the videos provided say that they don't supplement much. One of them points out that the pigs will chew the charcoal off branches and logs that have been in fires for the minerals. He deliberately puts out charcoaled wood pieces where his pigs are foraging for their culinary entertainment.[87]

Many homesteaders using pigs as part of their pasture and woodland management have a mix of woodland and pasture. They end up rotating them through the seasons from one to another and including some time in a permanent barnyard as well. Luke Groce explained how he rotates his pigs around his farm between pastures and woodland seasonally. For example, the pigs go in and clean up the pasture where the winter squash has been grown at the end of the season.

Humans have demonstrated that it is not only viable but advantageous to use your pigs as partners on a homestead in pastures and woodlands. Remember that you will not only want to keep the pigs in with secure fencing but also *keep predators out*. Your pigs will need protection. Read the following section about predators to decide how you will protect your pigs.

Sealing a pond

If you have a pond that needs some extra sealing for maintenance, or if you want to build a new pond, pigs can help!

Natural pond sealing has traditionally been done with *gley*, defined as *"a sticky clay soil or soil layer formed under the surface of some waterlogged soils."*[88] The modern farm community has turned this into a verb and uses the word "gleying" to describe the process of sealing a pond with clay (usually Bentonite) or livestock (such as pigs) to seal the bottom.

The theory is that the pigs compress the soil with their hooves and wallowing as well as adding manure to the pond. The result is both compacting and sealing.

Gleying was developed in the USSR by using manure to cover the bottom and sides of the pond and then layering straw or cut grasses with the manure, covering the manure entirely with the vegetable matter each time. After tamping it down, you wait 2 to 3 weeks and then fill the pond.[89] Apparently, they used this technique in building dams also.

It is vital to understand that letting pigs wallow alone will not seal up a new pond with a rocky bottom. There needs to be clay as a base, and they will help with leaks or enhance the seal of the clay.

There are success stories about sealing ponds with pigs. Farmhacker.com offers a video showing before, during, and after footage of the pigs helping to seal the pond.[90]

The best environment for this technique is where there is a hot summer. The pigs will **love** wallowing in the pond when it's hot. If you are starting a pond, they will start in the middle and work outwards as water fills the bottom.

Gleying with pigs is best suited to ponds that are wide with gentle slopes that are comfy for lying on. Your pond can be fairly deep, but the deeper it is, the larger it needs to be to accommodate happy pigs who want to bask in the sun while lying in the water on the edges.

Another happy farmer shows us the results and his process of sealing a pond with his pigs.[92]

From the blog "The Ozark House."
"Sealing a Pond With Pigs"[93]

It may take a couple of summer cycles for the pigs to get your pond in shape. You have to assess your pond's depth, width, and substrate, then customize your sealing plans to accommodate the time it will take to get it to hold water.

A county extension agent told one homesteader that if you are starting a new pond, it is recommended that you lay down bentonite clay first, then use the pigs to enhance the seal. The homesteader posted about it on a forum.[94]

My attempt at a pig pond failed for a reason. The hole in the ground was made in the 1980s by the previous owner of our property as a bunker. He used the technique of burying tires in the "walls" of the bunker to support it. There were more than 60 tires buried.

We pulled them out in the spring when the ground was still wet (early April) then we let the earth settle for more than a year. Then, when we got to July of the following year, we put some clay in the bottom of the hole, added water, and made a pig enclosure.

They would not go near the wallowing hole we made them. It was 90+ degrees, and they placed themselves around the edge and never got themselves wet. We put them back in the shade under trees in the forest and made a wallow for them, and they went back to lollygagging around in the mud.

Pigs have an incredibly acute sense of smell. Our only explanation for their rejection of the pond was that they knew it was still toxic (even after 40 years plus more than a year of airing).

We are now growing mullein and mushrooms to do the toxic cleanup. We will test the soil every year for toxicity and then try pigs again to get the job done or maintain the edges in the future, as demonstrated in the Ozark blog post discussed above.

Use of your Pig Sealed Pond

- **Irrigation -** After the pigs are removed for an entire year of all 4 seasons, the pond can be used as irrigation for vegetables. You don't want to use the water for vegetables that are grown for human consumption any more than you would put raw manure on your vegetable garden.

- **Livestock Water -** If you use the pond for livestock, talk to your vet for advice about when it is safe to use the water for the particular livestock you have.

- **Swimming Hole -** I never want to swim in a pond that had been lined with manure, would you? Just use bentonite clay.

- **Drinking-Water -** Despite the USSR lining their dam reservoirs with manure, I would not choose to assign the water from a pig pond as drinking water for humans, ever.

Conclusion of Pigs Working Land on a Homestead

We have discussed pasture, rototilling, forest feeding, and pond sealing. Consider your land and think about how you would plan for pigs as homestead partners.

- If you have a small homestead, you may want to have a couple of pigs who work the areas you need clearing or rototilling, then slaughter them for meat and wait till you have jobs for more.

- If you have a medium-size farm with large gardens for vegetables, pasture, and woodland, then learn from the interview with Luke Groce about rotating the pigs through the seasons.

- If you want to repair a pond, think about how long it will take to be able to use the water for your intended purpose. You can choose to let the pond function as a summer spa for your pigs while they help seal it for you, or you can use the manure to line it in layers as described.

- Whatever your intentions, planning is foundational, and then you can make the most of the pigs who will also provide you with amazing meat.

Truffle Hogs

History of Truffle Hogs

The first documentation we have of the use of hogs for truffle hunting was in 15th century. An Italian gastronomist, Bartolomeo Platina, mentions it in one of his books.[95] Although this is the first clear written evidence of it, academics believe that the practice was established during the Roman Empire. In 1985 Italy banned the use of hogs for truffling because of the damage they did to ecosystems, including the truffles themselves (thereby lowering production).

Why Hogs are So Good at Truffle Hunting

Hogs and pigs have a remarkable sense of smell. For example, they can smell truffles that are as deep as three feet underground. They also love to eat and are highly food motivated. But there's more...

In the case of the truffle, the mushroom has a hormone in it, *androstanol*, which is the same as the male sex hormone found in the boar's saliva.[96] Sows go crazy over truffles and will eagerly use their sense of smell to find them.

How you Truffle Hunt with Your Hogs

Even if you decide not to use a pig for truffle hunting, there may be a pig in your future who is exceptionally social and one you may decide to keep as a pet. This was true for me as well as the kids of my pig mentors next door. We decided to try truffle hunting for fun because it comes so naturally to pigs, and humans have had success for hundreds, if not thousands, of years. These games were fun for the pig, fun for us, and a great project for the children as they learned to train and care for animals.

First, use a female. This is true whether you are seeking truffles or using it as a game for a pet pig. Boars do not respond to the smell of the chemical in the truffles and are not motivated by it the same way that the females are.

For dedicated truffle hunting (vs fun with a pet pig), start training at 6 weeks to get them to recognize the smell of truffles, learn how to sit and walk on a leash. The following training instructions are based on my experience as well as the booklet, *How to Train a Truffle Pig*, by Emma Liles.[97] My experience was with Petunia, a pig I had decided was to be a pet. She was older than 6 weeks, (starting early helps the process, I'm sure; especially if you are training for serious truffle hunting), but at 11.5 months, she was young enough to be flexible, responsive, and learn quickly.

- **Sit**
  Ms. Liles uses a common dog training technique to teach a pig to sit.
  o Say "sit" and hold a treat over the head of the pig so that they must look up. This encourages them to sit in order to be able to see the treat.
  o She notes that you need to let the pig sit themselves, *not* push their hind down. Be patient, keep saying the command (firmly but kindly), and the pig will learn.
  o When she sits on command, give lots of praise as you would a dog, as well as a highly desirable treat.
- **Stay**
  o Your pig needs to learn to "stay" or "wait." You may have already done this when training your pigs to wait for their feeding. (See the training section.) If not, here is how to train one pig to wait.
  o Make her sit and say the command you choose to use for stay (use one consistently) - either "stay" or "wait."
  o Back up, repeating the command, then "OK!" to release her and give her lots of praise and a treat.

- For truffle pigs, progressions include lessening the repeats and turning away and walking a few steps, then making your distance and time a little longer.
- Remember that patience and incremental steps are the keys to success.
- 5 minutes a couple of times a day until she is consistent, and you can practice it randomly a couple of times a day; if the experience is positive, she will look forward to it.

- **Walking On a Leash**
  - You do not want to be pulled by a 200+ lb. animal; she will win. Start as early as you can and train your pig to politely walk with you and "heel" by your side.
  - Ms. Liles instructs that if the pig is walking on your right, hold the leash in your left and keep a treat in your right hand. This is an excellent idea. I held the leash with both hands across my body and had treats in my right pocket next to the pig. My method worked but delayed the response time with the treat when Petunia was obeying the commands. If it's in your hand, you can respond within a second to reinforce the behavior you want.
  - Get your pig to sit next to you, then say: "[Pig's name]...heel."
  - When you are walking, *never* pull on the leash; this will train her that she can control the situation with her weight and strength. Instead, change directions sharply and encourage her to follow you.
  - Training a pig to walk on a leash takes patience when you start. When I started with Petunia, we could only have a couple of sessions a day that lasted less than 10 minutes. After 4 days, there was a sudden breakthrough and it started to be a part of game time in her mind. She understood that she got her favorite treats. I had her full attention, and she just needed to learn what I wanted her to do.

- **"Drop it" Game**
  - I never saw any instructions about this, so I made it up. Getting a pig to drop food is asking something of them that is counterintuitive, to say the least. Even so, when they are happy and well-fed and bonded with you, it is possible to get past this hurdle. It seemed to me that taking a pig out to dig and find food and then trying to pry it out of its mouth without training was not a good idea, so I tried this.
  - I started with a non-food item - a tennis ball. I rolled it, Petunia ran after it, and I said: "drop it" (assertive, but with a perky, cheerful, positive voice, not a voice that would make her feel like she had done something wrong).
  - It took a couple of weeks for her to understand, but with treats and praise, she got it. Then, I started to mix up the objects, and she began to play fetch like a dog.
  - Once you are sure your pig understands "drop it" with non-food items, begin the challenge with food.
  - Get a bag of cheap mushrooms.

- o Make your pig sit and stay.
- o Put a mushroom in front of her (have a handful of mushrooms or treats behind you, ready to go).
- o Say "ok" then when she gets it, "drop it."
- o This took another couple of weeks. Petunia was gentle, but it was hard to get her to drop food. However, she was also smart and realizing that when I asked her to drop one thing, and it was replaced with *more* and *better* treats, she caught on.
- **Fetch**
  - o Start with a non-food item so that she knows that "fetch" and "drop it" will result in yummy goodness. A large plastic dog toy is good (*not* a Nylabone that has infused meat juices, you want it to be a non-food item).
  - o Progress from throwing a toy to tossing a mushroom, but she must bring it back uneaten to get the bag of mushrooms and treats you have.
  - o Patience is required; stop if you get annoyed.
- **Find the Mushroom Games**
  - o When you have at least some consistent success with "drop it," you are ready to advance to finding mushrooms. This worked well for Petunia; I made it up based on dog games. Emma Liles offers a similar game for pigs.[98] My theory was to **start** with finding mushrooms without paying for truffles. She happened to love them, so that worked in my favor.
  - o <u>Mushroom in a box</u>
    - ▪ You will need 5-6 paper boxes from packaging.
    - ▪ Set up 3 boxes and fill one with about 2 cups of mushrooms or less if the box is smaller. (Have extra boxes ready, she is likely to be enthusiastic about finding them!)
    - ▪ Tell her to "sit" and "wait" or "stay," then say [Pig's name]. "Fetch." She will learn this is her command to find hidden mushrooms, eventually in the ground.
    - ▪ Let your pig find the mushroom, help her open the box if needed.
    - ▪ Make her drop it, give lots of praise, and twice as many mushrooms as she found, plus a favorite treat.
    - ▪ Replace the used box with a clean one and put 1 cup of mushrooms in, repeat finding, drop it and treat reward.
    - ▪ Replace the used box with a clean one and put 2 mushrooms in, repeat finding, drop it and treat reward.
    - ▪ Lastly, replace the used box with a clean one and put 1 mushroom in.
    - ▪ When your pig can reliably find one mushroom in a box, graduate to hiding them in the ground as follows.

- o Mushroom in the Ground
  - This is the same idea but grab a pair of gloves and a garden trowel along with your mushrooms. I asked for help from a family neighbor for this one because I couldn't hide the mushrooms without her seeing them. (What a clever girl.)
  - Start with hiding several sets of 2 mushrooms together in one place, digging about 3" deep to start. Cover with soil and disguise the digging spot.
  - Give your pig a mushroom as a treat, make her sit and wait, then say "[Pig's name] ... Hunt!" or "Fetch!"
  - When she finds them, "drop it" and replace it with a handful of mushrooms with some treats as well. The reward for dropping it is **always** that she gets more.
- **Putting it All Together**
  - o When you've made some progress, then you can start hiding mushrooms in a forest landscape and taking your pig out on the leash to find them. If this is your own forested land, then your only concern is to ensure that you do not harm the roots of the trees with your digging and the rooting of your pig. If you are on public land, make sure that you know all laws and are not in a highly sensitive ecological area that could suffer damage from your truffle hunt practice.
  - o Hide just a few mushrooms; start with 4 or 5. Next, walk her on the leash using your commands and incorporate "fetch." Follow her, but do not let her pull on the leash.
  - o When she finds a mushroom, give tons of praise, and as above, more mushrooms with a couple of treats to reinforce that releasing the mushroom will result in something better.

I decided to add one more step as an experiment. I purchased truffle oils[99] and used a paper towel to put a thin coat on each mushroom when I hid it. *Use nitrile gloves so that your fingers do not smell like truffles!*

Petunia did seem more excited, I did not want to disappoint her after dropping them, so I had coated the treat mushrooms ready at hand as well before going out.

You are not going to be able to completely hide the smell of truffle oil from a pig whose sense of smell can locate a truffle 3 feet underground. However, you can minimize it. It took attention, but it was pretty easy to ensure that my hands were clean, no truffle oil was on my clothes, and the mushrooms for treats were in the tightest plastic containers I could find. The containers had to be washed thoroughly on the outside so that there was a little truffle smell as possible.

Do You Want to Truffle Hunt with Your Hogs?

As discussed earlier, using pigs for truffle hunting has been a tradition for centuries at least, perhaps millennia. Using mushroom hunting as part of your training and games for a chosen pet pig is one thing. Being serious about acquiring truffles to sell is quite another.

If you are serious about truffle hunting for its own sake, know that most people nowadays use dogs. Apparently, French truffle hunters using pigs tended to be missing parts of fingers.

Pigs can stress a forest by digging in the roots of the trees for the truffles. They are also heavier than the dogs used for truffling and have hooves instead of paws. Italy has outlawed the use of pigs for truffle hunting because of the damage caused.[100]

Conclusion of Truffle Hunting

I never actually went truffle hunting, I was just curious about the tradition, and since I found myself with a pet pig I thought I would try it. Petunia was pretty good at dropping the mushroom, I'm sure she could have found truffles deep in the ground, but it still took a lot of patience and she was not consistent with the dropping. She did it about 80% of the time. Nevertheless, it was an entertaining game that made her happy.

Even the small breeds of heritage pigs are heavy animals that are not inclined to travel, whereas a dog can jump in your car or go in a crate if crate trained.

Those who favor pigs for truffle hunting argue that even though dogs have enough sense of smell to find truffles, pigs are more motivated and excited about finding them because of the sex hormone contained in the truffles.

France still continues the practice of truffle hunting with pigs; there is even a competition every year.[101] But even in France, dogs have replaced pigs for most truffle hunters.

My advice is to use truffle hunting practice as an engaging, fun game with a special pig. It can also be bonding and a wonderful game for show pigs. If you want to get into truffle hunting seriously, do your research. Investing in a trip to France would be a wise idea as a business expense to learn from the experts. Follow the links in this footnote to get started.[102] Note that most truffle tours are about dogs, not pigs.

Showing

The homesteaders I know who are interested in showing have children who have raised pigs. Showing is an engaging way to allow a child 10+ years old to give specific care and attention to an animal and then get credit from a source outside of the family.

Two 4H-ers demonstrating excellent show pig skills[103]

Many adults enjoy showing pigs as well. It is a rewarding adventure and can build a reputation for the pigs on your farm if you want to breed and sell them at premium prices.

I have not shown my own pigs, but I have accompanied a neighbor's child to shows, and I helped with the training by *allowing the child to teach me* about showing and what needed to be done with the pig every day. I did my research beforehand and spoke to a couple of local 4H pig show experts so that I had context. Mark (my 11-year-old neighbor) advanced quickly in his skills as having to explain to someone else reinforced his own learning and built his pride and confidence going into the ring.

Some homesteaders get into pig showing because they happen to see a great show pig when they are purchasing from a breeder. Consider this option and learn how to spot a potential show pig.

Choosing a Show Pig

As with any show animal, you want a fine specimen of the breed. For pigs, the bone structure is the foundational consideration. Muscular development depends on genetics and diet. So if you get a pig who is in a lineage of winners, the bone structure looks solid, and the facial features are good, then it is likely that you will have an excellent show pig.

Finding a breeder who has pedigree pigs with documented lineage is the best way to ensure that the young pig you are looking at will grow into a winning show pig. If you are less serious about your pig showing results, then asking the right questions of the seller and carefully observing the young pig will give you a fair chance to train a winning pig.

One expert, Blaine Rodgers, emphasizes that choosing a young pig for its musculature is not the way to go. Bones are their structural foundation, so consider these points:

- Are they square in their build?
- Are the size and shape of their feet breed standard?
- How do they walk? Is their weight distribution even, or do they favor one side or one leg? [104]

A skinny young pig can become a winning show pig with the proper nutrition and exercise if the bone structure is solid. Keep this in mind, and you might get a great deal on a winning pig.

Let's discuss the specific breed standard characteristics to watch for in both American Guinea Hogs and Kunekune pigs.

American Guinea Hogs
(The following information is from the American Guinea Hog Association.)[105]
- Length: 46-56" (length is measured from the point between the ears to the base of the tail)
- Height: 22-27"
- Facial Structure and characteristics: AGHs have a "slightly dished" face[106] Their ears are medium-sized and upright. They may be a little floppy and bent only at the tips as adults. If you see a piglet with floppy ears, they would not be suitable for breeding or showing.
- Color: Most AGHs are all black. It is still within the breed standard criteria to have bits of white on the feet or end of the nose. Larger amounts of white should not be bred and will not do well in the show ring.
- The tail is upright and has one curl.

Kunekune Pigs
The American Kunekune Pig Registry has "sanctioned shows" with specific eligibility requirements. They also publish the judging criteria for various ages and groups. If you want to show a Kunekune pig, start here.[107]

The published breed standard for Kunekune pigs is more detailed than for the AGHs. Below is a general list, go to this site to get the details that include the points assigned to each characteristic. [108]

- Face is "broad and dished," teeth suitable for grazing, eyes not obstructed by the ears.
- Unlike the AGHs, the ears on the Kunekune flop and tilt forward.
- Neck short to medium in length, jowls are light to medium.
- Chest is a little wide with shoulders level.
- Back is level or may be a little arched.
- Legs are straight and able to support the weight of the pig.
- Colors must match one of the recognized color charts of the AKKPS. [109]

The list above includes a few examples on the complete list. I would recommend printing this out or having this page handy on your phone if you are watching for signs of a show pig amongst your Kunekunes, then you can observe and learn the criteria. Furthermore, if you intend to purchase a show pig, studying this list and bringing it to shows will give you a foundation for speaking to breeders and Kunekune show trainers and participants.

If you consider these lists, you can understand that it may be difficult to determine whether a young pig or piglet will grow into the desired characteristics. This is why lineage and genetic information are vital for choosing a pig to raise as a show pig. Here is a video about choosing a Kunekune pig that is breed standard. [110]

Shelter

A show pig will not be participating in the homestead grazing and rototilling. They need a very clean pen and, as we will discuss, will require special feed and attention. Even if you only want to train one show pig, if you have another pig to live in the special pen with your show pig, you'll have a happier and well-adjusted show pig.

The Utah State University Extension advises that you prepare your show pig enclosure by disinfecting the fencing, water tank (and nipple if it is not new), and feeder with 1 part bleach to 4 parts water. This practice ensures that harmful bacteria are not going to make your pig sick on arrival. [111]

The show pig farmers I know have used thick straw bedding for their enclosures. However, I've read that some use sand instead because this requires more muscular effort for the pig to move around. The downside of this is that sand will be much more difficult to keep clean as it will be heavier to shovel.

I always recommend a water tank with a nipple for all pigs, but it is a much stronger recommendation in the case of show pigs. Ensure that the nipple is high enough for the pig to reach easily; you will probably need to put the tank on a stand to get the right height.

Bringing your Show Pig Home

The strongest recommendation is to purchase your show pig from a reputable breeder who has shown to have excellent biosecurity practices on their farm (e.g., you should have had to wear plastic covers over your shoes to go in the barnyard and been given instructions to wear clothes that had not been in the barnyard with your pigs). It is still advised that they be strictly quarantined from other pigs as per usual.

Feeding

Young pigs are still building their immune system. Some traditional pig show experts recommend medicated feed if your show pig has come from an auction or another form of sale. [112] Often, homesteaders are not likely to want to feed antibiotics to their pigs. Kansas State University offers an excellent, science-based article on natural alternatives.[113]

Here is a short sample list of alternatives discussed in the Kansas State University publication:

- For growth promotion:
  - Feed with added enzymes
  - Probiotics
  - Antimicrobial peptides
- For disease prevention:
  - Immune modulators
  - Bacteriophages, endolysins, and hydrolases
  - Farm management and biosecurity

Amount of feed:

If you have a young pig, you'll probably need about 25 lbs. of pig feed each week. You want to choose the highest quality possible for a show pig. [114]

Shows have specific requirements for weight. A traditional hog show requires that pigs weigh 230 lbs. to enter. The Kunekune Sanctioned Show Rules do not list any minimum weight. If you are showing an American Guinea Hog, check with the individual show you are registering with. Some shows have an "underweight" class. This class may or may not be the right choice for your pig, so find out how the show accommodates smaller heritage breeds.

Whether or not your heritage pig has to weigh the standard 230 lbs., you will need to be weighing your pig and watching for any feed adjustments you need to make in the feed starting 8-weeks prior to the show. You want your pig to be an optimal weight, not more or less.

The critical importance of water cannot be emphasized enough. We have discussed the possibility of salt poisoning due to dehydration, but that extreme is not the only reason that water is important. Clean, plentiful water will encourage the pig to eat well and is also required for healthy growth.[115]

It is better to use a pig feeder rather than an open trough as it will be much easier to keep the pen clean. Here is a sample feeder.

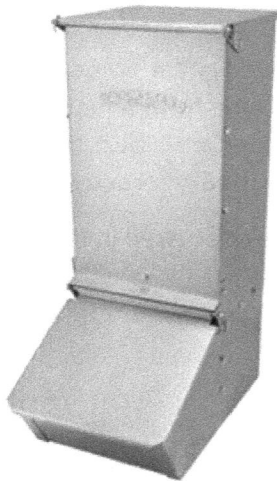

This image is from hogslat.com "Single Door Hog Feeder"[116]

Hot temperatures can lower the appetite of pigs. When it gets hot, give your show pig a nice cool rinse with a garden hose attachment on "shower." If the temperature is above 90 degrees for more than 6 hours a day, then do it twice a day. You want your pig to keep up their steady, healthy diet.

Exercise

Show pigs need exercise to build muscle. You will be combining their exercise routine with their training, of course. By the time you are 8-weeks before the show, your pig needs to be walking ½ mile, at least 3 times a week. [117] The 4H kids and adult pig show participants I've spoken to have walked their pig a minimum of every other day and usually every day. To give the pig stimulation and interaction, it is best to give them 2 walks a day, with one of them being ½ mile.

As with any animal, work up to the goal; in this case, ½ mile. Do not try to take your new young pig for a ½ mile walk the first day. Start small, make the walks something that the pig will look forward to. They are crucial as a point of bonding with you and training for the ring.

Handling

Show pigs need to be touched and handled a lot every day. In the ring, they will need to be examined and touched. Pet them, scratch them, examine their hooves, give a couple of treats, use touch as part of your bonding, and trust rapport with them.

Wiping Off When Exiting the Pen

Whenever you take them out of their pen, wipe off their snout, belly, legs, body, and under their tail. Speak gently and softly, but also be assertive and consistent. Start small, increase gradually. *We must be smarter than our pigs*! If your pig protests, you *must* give the impression that you are stopping *on your own terms,* not because the pig dictated. You must be the top pig. Show pigs have specific requirements but see the training section for more detail.

Grooming

Brush your pig gently every other day to keep their coat in the best health. Watch for any signs of dryness or other skin abnormalities. For training purposes, you might do it lightly every day. The principles discussed for wiping them off apply to brushing as well. It's good to start with a little and work up to a full brushing.

Your pig needs to learn to be cooperative for washing. If you purchase your pig when it is hot, start by giving them a gentle shower, add in scrubbing bit by bit until you can wash and rinse the whole pig. On hot days this will be easier as they will welcome it naturally. In cold weather, just wipe them off with wet rags and dry them thoroughly. If you purchase your pig in the winter or during chilly weather, start with the rags and wait for the shower until the temperatures get hot to be enjoyable for your pig.

Training a Show Pig

Leash Training

Your pig is not leashed while in the ring but training them to walk on a leash or harness is great for dominance as well as getting them to and from areas when you go to a show. See the training section for details.

Besides being trained to be handled and groomed, show pigs must learn the routines for the ring. The generic training detailed in the Training section of this book will build a solid foundation for the unique training required for showing. For example, your pig must not be pushy, argumentative, over-excited about their food, or aggressive. Instead, they must be polite, in control, and looking to you for safety and instructions when at a show.

Working up to Off-Leash Training

Assuming you've done basic training, including some leash work, you can take your pig into an open enclosure that is larger than their pen. (Preferably as large as a show ring, this varies by show.) When taken to a larger space, the pig is likely to run around. I would advise walking your pig on a leash in the pen once or twice a day so that the experience is no longer exciting. After it seems that your pig is accustomed to the pen, then let them off the lead and give them permission to run around if they want to.

When they have let off steam, call them to you for a treat, get them to sit, and begin your show training. Although you'll see in this video that pigs in a show will often run around when let out of the gate, it's quite entertaining and clearly a challenge for the less experienced Class 1 contestants.[118] Training your pig to come back to you after a bit of a run around is first. Ultimately, if you can train your pig to come out of the gate and stay with you, you'll be on your way to some prizes.

Guiding your Pig

In pig showing, the handler uses a "crop" or "whip" or "pipe" or "pole" to guide them. This list sounds like animal cruelty but is not a cruel device, and the pig is **never** hit or whipped. (In fact, one thing the judges look for is the handler's gentleness with the pig. Hitting your pig will get you thrown out.) The pipes and poles are simply straight, now usually made of PVC.

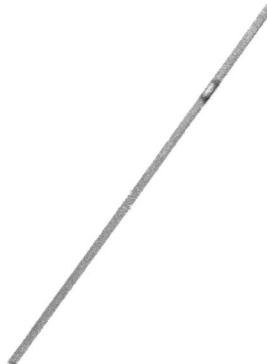

Traditional Clover Green Pig Pipe at Showstopper.com [119]

The whips have a soft end on them to get the pig's attention and help keep the head up.

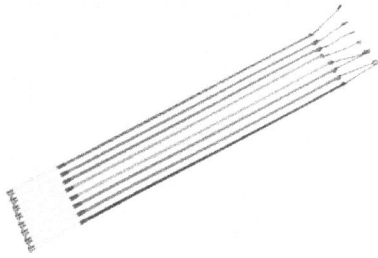

"Hog Show Whips"   "DW Whip Series"   "Easy Touch Pig Whip"
All three of these whip options are available at USWhip.com[120]

Many handlers prefer the whip because it is easier to keep the pig close to your body and remind them to keep their heads up. The head is supposed to be level with the shoulders as they walk around. This is counterintuitive for a pig as their nature is to be keeping that snout pointed towards the ground in order to smell.

Which you choose is a matter of personal preference. I was lucky to get to try a number of options because Mark's family had the tools. I preferred the whip to the pole. I did not like the whip with the tassel at the end like the "easy touch pig whip" (it was too floppy) but could use any of the other whips shown (the "DW whip series" or" Hog show whips").

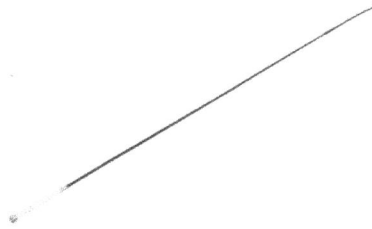

The basic idea is that you use the pole or whip to guide them by, for example, touching them on the left if you want to turn right. There are excellent short articles about using pig poles and whips and videos with explanations as well.[121] There is a specific article about using the whip to keep your show pig from running around in the ring.[122]

To become well trained yourself, read and watch videos, then go to as many pig shows as you can to observe the handlers. Using this method, you can decide which tool you want to start with and observe the instructions you've read and watched on videos in live action by both experts and those with less experience making beginner mistakes.

Keeping Show Pigs AND Humans Healthy

The pillars of health for show pig that we will discuss are:
- **Vaccinations**
- **De-Worming**
- **Guidelines for Human Health and Safety at Pig Shows**
- **Transitioning the pig from the breeder to your farm**

Vaccinations

Because of increased exposure from numerous farms, show pigs need to have a full set of vaccinations and de-worming. Note that it is not recommended to give them antibiotics as this is not effective against viruses and parasites that are the main problems. Use antibiotics if your vet prescribes it for an infection, not as a preventative tool.[123]

Note that pigs can experience vaccination effects just like humans can, so the vaccine protocol must be implemented well before the show. You don't want to stress your pig with having to be transported and show when they are weakened and don't feel well. Also, you only want to vaccinate healthy pigs – if they are ill, wait until they are fully recovered.[124]

According to the Texas A&M AgriLife Extension, a standard, basic vaccination routine is fine for your show pigs. The following is a simple summary; please read this article for details and options for various ages.[125]

- Actinobacillus pleuropneumonia (APP).
  - APP causes illness and sudden death.
- Erysipelas
  - Erysipelas causes skin disease and lameness and can result in sudden death.

There is a combination vaccination for these two diseases. The vaccine can be given as early as 4 weeks with a booster shot 2-3 weeks later. The piglet will be immune until 6 mos. of age. Consult your vet about further boosters.

- Porcine Reproductive and Respiratory Syndrome (PRRS).
  - This disease causes pneumonia, death, slows weight gain, and reproductive dysfunction.

The vaccine options for PRRS include only the respiratory aspects as well as others that include the reproductive and weight loss symptoms. One form of it can be given as early as 3 weeks of age and no booster is required. Read the details to understand, then consult your vet.[126]

- Mycoplasma Hyopneumoniae
  - This is another respiratory condition that results in pneumonia.

There is a routine vaccination given by at least 3 weeks old. It is one dose and will last for 6 months.

You can see that a reputable and experienced breeder with papers to show the vaccines given to piglets is a big plus for any pig, but especially a show pig who will be exposed to many other pigs on a semi-regular basis.

Some homesteaders want to administer vaccines themselves as a cost-saving practice. If you choose to do this, we recommend that you have a vet visit to explain it, demonstrate giving a vaccine to a pig, and give you the details for the boosters.

De-worming

Like all new pigs, your show pig should be de-wormed upon arrival to your farm and then again 1 month after before being released from quarantine. Ask your vet about the routine they suggest for maintenance.

Guidelines for Human Health Safety at Pig Shows

Swine flu is transmitted from pigs to humans. It spreads via contact with surfaces and is also airborne. The CDC offers an article specifically about protocols for protecting yourself and your family at shows.[127]
The basic guidelines are common sense:
- Don't touch your mouth or eyes when in the barn with pigs
- Wash or sanitize your hands thoroughly after being in the pig barn.
- Do not eat or drink while in the pig barn.
- Do not bring young children, strollers, pacifiers, bottles, or toys into the pig barn

Read more for the details.[128]

Transitioning Your Show Pig to Your Farm from the Breeder ad After a Show

Here is a quick reminder of the guidelines outlined in this book:
- Purchase from a reputable breeder, preferably a pig who has been raised from birth on one farm.
- Quarantine your pig when they arrive at your farm.
- Make sure they have an abundance of clean, fresh water. If they have been drinking from a trough when you got them, they may not understand how to use a nipple waterer. A trick is to wedge a kernel of corn in the nipple so that the pig smells it and, when exploring, discovers the water. Keep a close eye on this; we have discussed the dangers of dehydration for pigs earlier.

- Transition the feed from the breeder gradually. Pigs can die of edema caused by a sudden change in diet.

See the sections on water and introducing new pigs for more details.

After a show, your show pig will need to be quarantined again to ensure that it is not bringing any disease to your herd. Biosecurity measures such as plastic covers on boots, not sharing equipment between pens (or sanitizing anything that goes from the show pig to the other pens) are critical. The show pig participants I know always quarantined and de-wormed their pigs after a show.

Conclusion of Show Pigs on Your Homestead

As discussed at the beginning of this section, pig showing is not for every homesteader. Understandably, many farmers just want to keep their healthy pigs on their land to do the rototilling and brush clearing. Pig showing is a great way to gain a reputation for your farm and meet experienced pig farmers. It is also wonderful training for middle school children and teenagers as they learn to pay attention to the details required for livestock care. The bond with a show pig that you have trained well from an early age is rewarding. Consider showing a pig as part of your homestead strategy and practice.

Training

The temperaments of American Guinea Hogs and Kunekune pigs are celebrated and regularly described as "gentle," "trainable," and "docile." This is true, *and* trainable is the operable word. The use of the word "pig" to describe humans who are pushy, impolite, and food-obsessed arose for a reason. I would watch my American Guinea Hogs lying close to each other every day, basking in the sun being sweet with each other, but when feeding time came, they would become pushy and even aggressive with each other. Then they were gentle and docile again.

If you do not train your pigs at all, you will have large muscular beasts who think they are in charge. They will be difficult to examine for health purposes and difficult to feed. Even if they aren't aggressive, pigs get excited and can step on your feet or knock you down.

Pigs are herd animals with a clear hierarchy. You need to be on top, and you want them to trust you. Never hit - you can speak strongly and assertively when needed, but yelling is not productive.

Before I had my own pigs, I had the experience of providing an overnight house sit for people who had three American Guinea Hogs. The pigs roamed free inside a large stockyard and pasture with cows. I went inside the stockyard at feeding time, pulling a cart that held their feed in a bucket.

They came running over excitedly, stepped on my feet, pushed over the cart, and nearly knocked me over. The food went all over the ground, and they started pushing and fighting with each other to get to it.

Fortunately, I was wearing my steel-toed work boots and was able to keep myself from getting pushed over. Nevertheless, I had two main takeaways in that experience:

1) Even small pigs are powerful and potentially dangerous animals – *even when they are not being aggressive*. I've not encountered an aggressive pig; I don't want to.
2) I decided that when I had pigs, I was going to figure out how to train them.

Pigs are *similar* to but not the same as dogs. Pigs are as (or even more) intelligent as dogs, but it is important to understand that they are not as driven to be eager to please and desire connection in the same way as many dogs.

Individuals vary, but as a rule, there is a difference in the motivation for training. My mentors all said that you could train a pig in a way *similar* to the way you train dogs. I've had a lot of dog training experience and mentorship, so I used a lot of those principles, bearing in mind that pigs are not dogs. I was happy with the results and have assisted a couple of other homesteaders in their pig leadership.

The people who hired me to house sit had not bothered to train their pigs; however, the person who fed them every day had acquired their respect over time so that they did not knock him over. Therefore, after speaking to a couple of mentors, I took these steps, and I recommend them to you:

1) Purchase piglets who are weaned but small enough to hold.

I was advised to pick them up and put them down as much as I could and hold them until they were still. I was warned that they would squeal and thrash (they did) but told to hold them firmly until they settled. This is critical. If you put them down because they squeal, you may as well sell that pig because you've just trained them that they are in charge. Pigs are overdramatic - the squealing and thrashing are "over the top." Do not give in, hold firmly, and speak calmly. Put them down when **you** are ready to put them down, not when they are throwing a tantrum.

While holding them, stay calm and kind. You are both gaining trust and showing you are the top pig. Keep this up as many times a day as you can until the piglet allows you to pick them up and does not protest. Then, after they have stopped running from you or protesting when you pick them up, do it least a few times a day until they are too big to be held any more.

2) Teach them to be polite at feeding time.

The general principle here is that the pigs learn that they get fed when they are quiet. Unlike dogs, treat rewarding is not useful because pigs will come to expect it and then get aggressive with you if you do not have the treat. In *any* situation, feeding or otherwise, they *never* get what they want from screaming.

When it was feeding time, I walked into their enclosure with their food bucket and made a hand signal like "stop" and said "wait" (firmly). I stood still. When they were quiet, they got food. I gave them just a little at first to repeat the routine and help them learn and catch on to what I wanted them to do. It did not take long for them to understand and be quiet pigs who got their entire feed at once when I was satisfied with their behavior.

3) Consider teaching them to walk on a leash or be trained like a show pig to be guided by a whip or pole.

These skills can be a worthwhile investment for homesteaders who are rotating their pigs from one pasture to another. It is also useful in getting them to another enclosure to see a vet or lead a sow to a farrowing enclosure. I once spoke to a homesteader who said that the leash training was critical when she had to evacuate the farm due to wildfire. If the pigs had been unmanageable, they would not have survived; as it was, they were loaded into a trailer quickly, and everyone evacuated safely.

Leash walking and show guiding are excellent practices for bonding and maintaining your status as the dominant pig, so these are highly recommended as beneficial activities.

There are some excellent resources for pig training on the internet. One I found is a woman who produced a very clear and helpful video about pig training. She uses the "pick them up as much as possible" practice as well as similar ways to get them to be polite pigs at feeding time.

She also offers ideas that I've not mentioned, such as stomping, clapping your hands, and saying "NO!" firmly, as well as using your hand to push on them as they push on each other to make corrections to their fellow pigs.

Adri Rachelle is worth watching - a true pig expert.

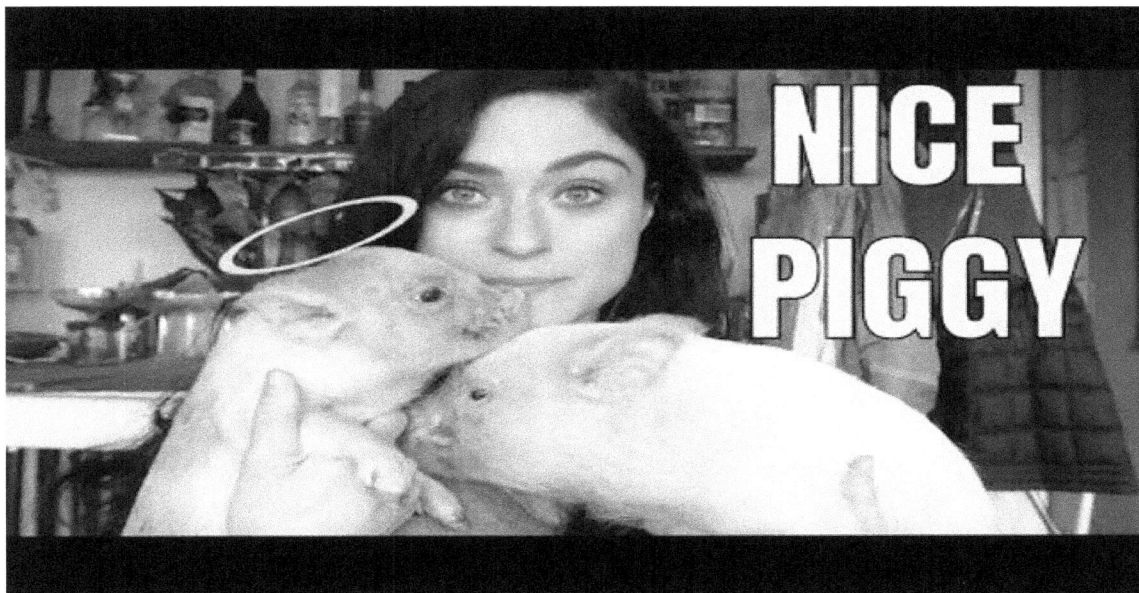

YouTube Channel "Adri Rachelle," video "How to Raise a Polite Pig"

Bored pigs are troublesome, so keep your pigs engaged and relational by handling them every day and making them be polite at feeding time. If you choose to leash and show train them, then you'll have more established dominance and useful skills as well. It takes time and patience, but a trained pig is an easier pig.

Predators

If your pigs are working pasture and forest, then it is likely that you will need to protect them from predators. The most common cause of fatalities is domestic dogs, but coyotes, wolves, mountain lions, and bears can also be a threat.

If you aren't watching your pigs all day and bringing them in at night or keeping them in a place where they will be predator-free, then you will probably need the assistance of a guardian livestock dog or donkey.

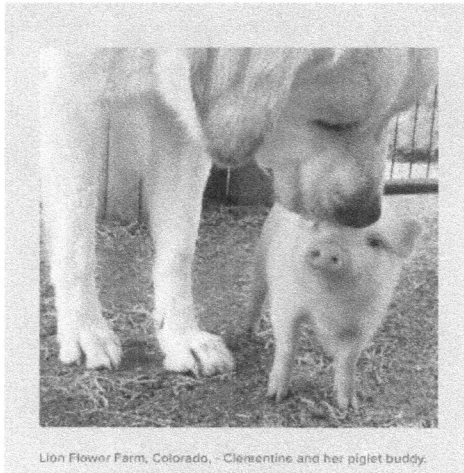

Lion Flower Farm, Colorado, - Clementine and her piglet buddy.

coloradomountaindogs.com
"CMDs in Photos" tab[129]

The first livestock guardian dog breed is traced to Mongolia. Over time, humans discovered that large, domesticated canines were a great asset to protecting herds of livestock.

There are many breeds of LGDs. Over four years, I have lived with three adults long-term and litters of puppies who were sold to be trained for homesteads. The people who share our homestead land are professional breeders and trainers of LGDs. The dogs who protect my pigs and goats are Great Pyrenees, Karaakachan, and one is a mix of the two. My strong preference as a breed is the Karaakachan. All of the dogs are full-time workers on the farm, and they are also habituated to the children on the homestead.

Another interesting breed is the Colorado Mountain Dog, which has been bred for 25 years with the focus on temperament and traits preferred by homesteaders in the 21st century (such as selective barking). The Colorado Mountain Dogs are also bred with the small acreage homestead in mind vs. large-farms or herding livestock over great distances. Here are some breeder resources to get you started.[130]

If you decide to acquire an LGD, you need to understand key points:
- LGDs are large (approximately 120 lbs.), and they eat a *lot*.

- They bark. One person said that trying to get your LGD not to bark is like taking the light bulb out of a security light. It's what they do.
- They are territorial and can be aggressive or even dangerous to other dogs. Fencing is key.
- The dog(s) need to be trained to accept anyone you introduce. You don't want your dog to be aggressive to the vet or other visitors.
- Get credible support and education before deciding on the breed that is best for you. Then, find a reputable breeder. There are forums and groups that include experienced LGD owners; they can be a rich source of information.
- Do a lot of research on the breeder. There are many out there who breed for impressive size. A super-sized LGD may have joint problems, and size is not the determining factor in effectiveness. Also, prioritizing size means that there may be ill-tempered dogs and lacking intelligence who are being bred.
- Find a trainer who has specialized in training livestock guardian dogs to protect a herd. Pay for it.
- Ideally, be in touch with a local person who has experience with LGDs on their homestead.

Other Livestock Guardian Animals

Donkeys and Llamas are also used to protect livestock. Llamas are generally suggested for sheep and goats; they might be suited to pigs if they can bond with them. Donkeys are naturally aggressive toward canines; they are fierce and known to kill coyotes who threaten a herd. Llamas are also dog-aggressive and have the plus side of not being prone to wander or test fencing. Both llamas and donkeys are also prey, however, and cannot be expected to protect against bears and mountain lions without the presence of an LGD. See this footnote for more details about donkeys and llamas as guardian animals.[131]

Conclusion of Livestock Guardian Animals

Our pigs stayed in our knapweed meadow for 18 months and were *never* threatened by the coyotes, bears, or mountain lions who regularly stroll across our property. The LGDs were not in the pen with them but having the dogs in the goat enclosure near the pigs was enough to keep the predators at bay. 100% effective security just being *near* was an impressive track record considering that before the LGDs were introduced, the local bears would regularly cross over the meadow in front of our house pictured in the previous section regarding rototilling.

Whatever your choice, know your predators and know what measures are recommended to protect your pigs in your specific location and homestead setup.

Part Four: Breeding & Butchering

Breeding Pigs

There are general guidelines about timing when breeding small herds of pigs. It is advised to think about the seasonal changes in temperature. The Penn State Extension advises that you consider these facts:

- Begin breeding in the spring. Piglets and young pigs will have the best growth in temperatures around 70 degrees F. They are fine with cold temperatures as they get older and gain weight, but farrowing during winter or early spring months is not ideal.
- When will your pigs reach market size? AGHs are mature in 6 months, Kunekunes are ready in 14-18 mos.
- Pigs that are full weight will eat more feed when they are in 50-70 degrees F. When it is over 70 degrees, they will eat less.[132]

In addition, as part of your breeding program, consult your veterinarian about vaccinations or deworming and what timing they recommend.

Breeding American Guinea Hogs

Age of Maturity

Boers: The AGHA says as early as 6-8 months[133]. I go with the latter figure to ensure the best health of the sperm. You want the sperm to be fully mature, healthy and, have the strongest virility.

Gilts: Can come into their first heat and be bred as early as 8 months. If bred at first estrus, they can farrow near their first birthday.

Gestation

114 days (three months, three weeks, and three days)

Farrowing

AGHs are cooperative with handling during pregnancy and even during labor if necessary. Typically, all goes well, and there is no need for human intervention.

Make sure that a farrowing space is provided for her. All she needs is clean, dry straw out of the elements. Make sure it is deep enough for her to dig into making herself at home.

Manage your herd so that sows are not over-bred. An AGH can have 2 litters per year.

Piglets

Litters can be 1-14 piglets, but the most common is 6-8. Our litters have all been 5-8.

Piglets are born with their eyes open, and they are "about the size of a soda-pop can with legs."[134] It is amazing to watch them be able to find the teat to feed immediately.

If you are going to castrate any of the males, this is usually done at 14 days.

AGH piglets wean at 5-8 weeks.

Breeding Kunekune Pigs

Choosing which pigs to breed

There is a difference between "line breeding" and "inbreeding." Linebreeding is choosing to breed pigs for desirable characteristics (often temperament). Inbreeding is when parents of pigs breed with their piglets or littermates breed with each other.

The Kunekune community has a lot of talk about "COI" (Coefficient of Inbreeding). These principles apply to American Guinea Hogs as well. Inbreeding can be a problem and cause genetic defects in the breed. The coefficient is the percentage of the genetic makeup shared with the parents. You want a *low* coefficient in your pigs. Inbreeding can cause:

- Lower body weight
- Inherited deformities and diseases
- Fertility problems and lower litter numbers
- Stillbirth
- Congenital disabilities
- The piglets can have lower vitality and resilience, making them more prone to illness

For more resources on COI, use these resources for context before speaking to your veterinarian and experienced pig breeders. [135]

Age of Maturity

Boars: They can breed at 8 months, but they are not fully developed. It is recommended that you wait for 1 year.

Gilts: At *least* 10 months of age; again, it is recommended that you wait till 1 year.[136]

Getting Pigs to Breed

Generally, this happens naturally and easily. The estrus and mating should be tracked; however, one boar should only copulate every 2 weeks to keep the sperm as healthy as possible.

When I had my first herd, I watched our boar mount a sow, and mid-copulation he stopped still. She looked back at him like "Dude...what *are* you doing?" Then he fainted! He just toppled over on the ground. Boars are not known to be romantic, but the sow clearly thought that was more than a little disappointing. She went over to him to nudge him; he was out cold. I wondered if he had died. About 3 minutes later, he got up and went back to his daily activities as though nothing had happened.

I spoke to our vet about this incident, and she thought that "Casanova" had probably been overdoing it, and needed to rest and recover. Fortunately, all of our piglets were healthy American Guinea Hog specimens. Still, after that incident I was more intentional about keeping the boar away from a sow or gilt in estrus when he had already been sexually active.

Gestation

116 days – like the AGHs, the standard measure is "three months, three weeks, three days."

Farrowing

About a week before expected birth, let the sow "nest" in her farrowing enclosure. As with the AGHs, she will want to have deep straw or pine shavings to make a depression and get comfortable.

Kunekune piglets may not be as cold-hardy as the AGH piglets. Ask your vet or local pig experts whether you may need a heat lamp to help ensure piglet survival if your sow is farrowing in the winter.

Piglets

Litters tend to be 6-8 piglets.

Kunekune piglets feed on the teat longer than AGHs. They are weaned at 8-10 weeks.

For either AGHs or Kunekunes, the spring is the best time for farrowing, rather than the extreme temperatures of either winter or summer. Think ahead about what your temperatures are expected to be and make provisions accordingly. Ensure that your farrowing enclosure is draft-free but well ventilated. Here is a guide from Washington State University about caring for piglets in the winter. It is written for market pigs but offers advice about calculating your pig's needs depending on size.[137]

Butchering Pigs

Butchering and processing are skills that should be taught face-to-face by an experienced mentor. This book offers some context, guidelines, and resources, but nothing can replace a human physically present who shows you once, then watches you do it the next time, being there for questions you have or any corrections to make.

Age to Slaughter

American Guinea Hogs are ready to butcher at 6-9 months. Calculating the carcass weight is important so that you have enough, but not too much, meat to store in your freezer. You can roughly say that you'll get 50% carcass weight as edible meat. For example, at 6 months, if your pig weighs 60 lbs., then you'll have 30 lbs. of meat to store. If you wait until the pig is 120 lbs., you'll have twice that much.[138]

Kunekunes are smaller, which can be advantage if you are just feeding two or three people. When they reach 22 pounds (approximately 14-18 months), you'll get 12 pounds of meat.

Seasons come into play as well as size when you are butchering. If you butcher in the early summer, they will have been feeding on the spring grass, which is full of protein. This makes the meat more flavorful. On the other hand, the younger the pig is, the more tender the meat will be. Think ahead and consider both age and the season in which the pig will be butchered.[139]

How to Slaughter

If you are going to sell any of your meat or give it away to a charity, you *must* have it slaughtered and processed in a USDA-approved meat processing plant. The following instruction guide applies to meat that will be in your freezer for your family.

The most common way to slaughter a pig is with a firearm. Done correctly, your pig will die instantly and not suffer. It is recommended that a 12, 16, or 20 bore shotgun be used. The right gun will help ensure that the bullet gets through the thick skull into the small brain. Here is an excellent article that includes anatomical (not graphic) illustrations as the author explains how it is done.[140]

I chose to read the resources that I list below from the Humane Slaughter Association, and then I paid my vet to come and demonstrate how to do it. I asked an experienced and trusted pig mentor to watch me as I did it on my second pig. They talked me through it again, and I was able to get it right. After that, I felt confident and was able to slaughter them without undue stress to either the pig or myself.

One practice that is of high importance to me is not to kill an animal in front of other animals. I make sure that other animals are penned away from sight near the A-frame that I use for hanging, where I slaughter my pigs.

The Humane Slaughter Association, based in the UK, offers training courses, including online training and .pdf documents, to assist farmers in learning how to slaughter livestock, including pigs.[141]

If you use a firearm to slaughter your pig, make sure that you have the required license or registration for your state.

If you are going to take your pigs to a butcher, you will need to haul them. This is another great reason for training. In fact, you could train your pigs to get in and out of the trailer, so they are already comfortable in it. The ride will be distressing enough, so make the loading go with as much ease as possible. According to Penn State Extension, pigs who are stressed in the hauling process produce lower quality meat.[142]

It is (wisely) suggested that your loading ramp have sides so that they can't change their minds and jump off. Let them go at their own pace, don't push, and be patient and kind.

States vary regarding regulations for health certificates when you bring pigs to a butcher, so ask the processing plant what the requirements are to avoid an unwelcome surprise.

Processing Pig Carcasses

If you are processing your pigs on your homestead for your family, you'll need to prepare. Even with the resources provided as this section describes the process, find a mentor who will go through the process with you on your first pig and then watch you do it the second time around.

Have a cart ready that will hold the weight of the pig. Once the pig is dead, you have a very heavy thing to lift and move to hang, and then get to your hanging spot and processing table.

The first steps include videos that are demonstrations by a couple who are processing a Kunekune pig. Next, the skinning and butchering into cuts are demonstrated by "The Bearded Butchers," who have an artisan pig business. They are all experienced, and you can see the steps:

1) Lift the carcass onto a cart and scrub the whole pig thoroughly; as demonstrated in this video,[143] I use a scrub brush and hose; they have a scrub brush as a hose attachment which looks like a great idea.

2) Poke holes in the back lower leg above the "ankle" as shown here[144] Note the two thick 9" nails are used. They do not wear nitrile gloves for the process. I recommend that you wear them and keep a box around so you can change them out if you get feces on your hands as you gut the pig.

3) Hang the pig as shown, starting at 6:30 with the tying of the knots on the nails. [145] Note the A-frame structure they have built as a multi-purpose tool. (I had one, which was fantastic, but a grumpy young bull destroyed it.) You should be able to get a hose to your hanging spot. Hoist the pig up to hang it.

4) Be prepared with a bucket and several heavy-duty "contractor" trash bags. See how he slits the pig down the center of the pig at 1:08; the head is then cut off. As they note, some people keep the livers of Kunekunes because they are said to be delicious. Other parts of the guts are disposed as food for dogs or trash.

5) Another wash down and scrub at 6:55[146]

6) Skin the pig. The Bearded Butchers demonstrate skinning with a much larger pig in their artisan meat processing plant. Even though your AGH or Kunekune will be much smaller, their demonstration of how to make the cuts so that the skin comes off easily applies to pigs of any size.[147]

7) Move the pig carcass from the hanging spot, back on the cart you used before, and take it to your cutting table.

For my first AGH pig butchering, I used a folding table with metal legs and a plastic top. It worked fine the first time, but the second time it broke mid-process, which was stressful. Those tables are not made for this! Take the time and investment to make a solid table at the right height for your back that can hold up to 300 lbs. Use a surface that can be sterilized easily. Ours is sanded wood with a non-toxic sealant. Set the space up with plenty of bright light; you'll need to see – barns can be dark!

8) The Bearded Butchers demonstrate and explain how to section your pig into the classic cuts.
[148]

Here is a PDF resource that illustrates the cuts on a pig. I studied this before my first butchering, and it helped a lot.[149]

Conclusion of Butchering and Processing

All of these butchering resources are for you to use as context and education so that *when* you work with a mentor, you can know more about what to expect, the tools you will need, and how the process will go. Trying to learn this kind of skill on video is "doable" in theory but is a setup for frustration and the stress of time pressure since you have a carcass that needs refrigeration. I strongly recommend working with a mentor to show you the first time. The second time they can watch you do it, make corrections, and be there for your questions.

Conclusion

American Guinea Hogs and Kunekune pigs can be a valuable, efficient, and enjoyable addition to a homestead. These smaller heritage breeds lend themselves to being easier on the land and easier to handle in any situation. They also do not require so much freezer space if you are raising them for meat.

Before getting your pigs, remember these points:
- Assess your land. How can pigs be partners in caring for your land as well as providing you with meat? Will they be eating mainly grass or brush clearing in the forest?
- How many pigs make sense for your land and meat use?
- How much feed will be required seasonally to support your pigs?

Ensure that your enclosure is secure. If you are using electric fencing on a pasture or wooded area, set it up before getting your pigs. You'll need to test it, because they will. Follow the instructions in this book for training a pig to respect the fence.

Decide on your supplies, such as feeders and waterers. Set them up before you get your pigs.

If your pigs are out on a pasture or in a forest, how will they be protected? If you intend to have a livestock guardian dog, then the electric fence will likely need to be taller, and they will need training to bond with your pig herd. Even if you are keeping your pigs in a barnyard near your house, you may need protection from large predators and dogs.

As discussed, pigs can help enhance the vitality and health of your land, but they need to be rotated so that they do not become destructive. So plan ahead and know where you will be moving them next.

Be the leader pig to them. Don't let them push you around and insist that they be handled easily and polite at feeding time. Train them to walk on a leash and go into a trailer if you plan to haul them to a butcher. Train them and play games as we've described to reinforce benevolent dominance and enable you to handle them and lead them where you want them to go.

Whatever uses your pigs give you, find mentors. Whether showing, rototilling, pond sealing, or meat, find local experienced people who can talk to you about your specific situation, land needs, and set up. Enjoy your pigs!

Endnotes

[1] https://smithmeadows.com/farm-products/grass-fed-pork/attachment/p4110063/

[2] https://www.permies.com/t/77500/Ultimate-Permaculture-Pig

[3] https://www.cdc.gov/flu/swineflu/exhibit-pigs-at-fairs.htm

[4] https://osbornelivestockequipment.com/news/how-creep-feeding-piglets-increases-yields/

[5] 10 Pig Breeds for the Homestead
https://www.iamcountryside.com/pigs/10-pig-breeds/

Fourteen Heritage Pig Breeds
https://bigpictureagriculture.blogspot.com/2015/11/fourteen-heritage-pig-breeds.html

Best Heritage Breeds of Pigs for the Homestead
https://thefarmerslamp.com/heritage-breeds-of-pigs/

Pig Breeds: A Handy Guide
https://www.reformationacres.com/2018/01/choosing-pig-breed.html

[6] https://guineahogs.org/life-cycle-of-american-guinea-hogs/

Payne, Cathy R. Saving the Guinea Hogs: The Recovery of an American Homestead Breed. Rose Garden Press, 2019.
- This is a must-read for anyone deciding to raise American Guinea Hogs. She includes interviews with farmers who grew up with these hogs in the 1940's, their stories are fascinating and incredibly useful for understanding how to raise these pigs and how they can help us on a small homestead.

[8] https://morningchores.com/kunekune-pigs/

[9] https://howtospecialist.com/outdoor/shed/pig-house-plans/

[10] "Tools for a Successful Pig Fence." I love this article because it explains the foundations for success
https://www.iamcountryside.com/pigs/successful-electric-pig-fence/

Beginners Guide to Electric Fencing for Pigs
https://kippax-farms.co.uk/pigs/electric-fencing-for-pigs-and-hogs-guide

[11] Electric Fencing Guide on the Tractor Supply website:
https://www.tractorsupply.com/out-here_articles_fencing_pig-hog-fencing

The 3 Best Pig Fences in 2021
https://farmhacker.com/best-pig-fences/

Options for electric netting rather than fencing for pigs:
https://www.premier1supplies.com/hogs_pigs/fencing.php

I love the chart on this site for spacing instructions
https://www.zarebasystems.com/learning-center/animal-selector/pigs-hogs

[12] https://www.offthegridnews.com/how-to-2/the-easiest-way-to-train-pigs-to-an-electric-fence/

[13] https://www.offthegridnews.com/how-to-2/the-easiest-way-to-train-pigs-to-an-electric-fence/

[14] Ibid.

[15] https://www.youtube.com/watch?v=CEeoHiNaTUU

[16] The Easiest Pig Waterer
https://www.youtube.com/watch?v=7x7ORJFnyn4

[17] https://www.youtube.com/watch?v=7x7ORJFnyn4

[18] https://agrilifeextension.tamu.edu/library/4-h-youth-development/keeping-show-pigs-healthy/

[19] https://www.amazon.com/Rubbermaid-Commercial-FG424300BLA-Structural-Capacity/dp/B000NPBLAU/ref=as_li_ss_tl?dchild=1&keywords=pig+waterer&qid=1586972833&sr=8-74&&linkCode=sl1&tag=farmhacker-20&linkId=5f41e9dffe8641422afbe0f5ea09d003&language=en_US

[20] https://www.amazon.com/Rubbermaid-Commercial-FG424300BLA-Structural-Capacity/dp/B000NPBLAU/ref=as_li_ss_tl?dchild=1&keywords=pig+waterer&qid=1586972833&sr=8-74&&linkCode=sl1&tag=farmhacker-20&linkId=5f41e9dffe8641422afbe0f5ea09d003&language=en_US

[21] https://www.amazon.com/Miller-Rubber-Feed-quart-Black/dp/B000BD8JYK/ref=sr_1_21?dchild=1&keywords=small+livestock+trough+round&qid=1618342830&sr=8-21

[23] https://www.valleyvet.com/c/livestock-supplies/equipment-supplies/livestock-equipment/swine.html

[24] Ars Hoof Trimming Shears

https://www.premier1supplies.com/p/ars-hoof-trimming-shears?msclkid=63e87ab2b93b1b84e0991a05647f8b42&utm_source=bing&utm_medium=cpc&utm_campaign=(ROI)%20Shopping%20-%20Clippers%2FShears&utm_term=4583657829015238&utm_content=Clippers%20%26%20Shears

25 https://healthypigs.com/product/hoof-trimmers/

26   https://www.qcsupply.com/540993-hoof-trimmer-hd.html?msclkid=0e933da19ce3117687095d1641a76e29&utm_source=bing&utm_medium=cpc&utm_campaign=Shopping&utm_term=4577541785812491&utm_content=All%20Products

27 Here is an Amazon page showing wire saws used by veterinarians
https://www.amazon.com/gigli-saw-wire/s?k=gigli+saw+wire

28 Here is an example of the work gloves I use – they have dexterity but the protection I want for pig purposes:
https://www.amazon.com/LANON-Chemical-Resistant-Reusable-Protection/dp/B089QHMZDX/ref=sr_1_4_sspa?dchild=1&keywords=rubber+work+gloves&qid=1618426707&sr=8-4-spons&psc=1&spLa=ZW5jcnlwdGVkUXVhbGlmaWVyPUEzOVJZNzdXVUpaRldNJmVuY3J5cHRlZElkPUEwMjgxNzcyN0RMTllDRldZVVRRJmVuY3J5cHRlZEFkSWQ9QTA4NDA2ODkyRzhYV0MyRlhFS1ozJndpZGdldE5hbWU9c3BfYXRmJmFjdGlvbj1jbGlja1JlZGlyZWN0JmRvTm90TG9nQ2xpY2s9dHJ1ZQ==

29https://thepaleomama.com/2017/04/10/american-guinea-hogs/#:~:text=AGH%20eat%20roughly%204%20percent,to%202%20gallons%20each%20day.

30 Water Recommendations and Systems for Swine.

https://extension.psu.edu/raising-small-groups-of-pigs

31 https://www.thepigsite.com/articles/common-causes-of-poisoning-in-pigs
32 https://opensanctuary.org/article/things-that-are-toxic-to-pigs/

33 https://sugarmtnfarm.com/animals/pigs/

34 https://sugarmtnfarm.com/animals/pigs/

35 This is a reliable, high-quality brand that offers mixes for piglets, sows and finishing
https://blueseal.com/product-lines/home-fresh/pigs/

Tractor Supply offers a range of pig feeds standard to premium:
https://www.tractorsupply.com/tsc/catalog/pig-feed

36 https://sugarmtnfarm.com/animals/pigs/

more detail: https://sugarmtnfarm.com/2011/10/03/rootless-in-vermont/

[37] https://www.roysfarm.com/feeding-pigs/

[38] *https://extension.psu.edu/raising-small-groups-of-pigs*

[39] Ibid.

[40] https://www.jlgreenfarm.com/forested-pork

[41] https://www.chlpi.org/wp-content/uploads/2013/12/Leftovers-for-Livestock_A-Legal-Guide_August-2016.pdf

[42] https://petpigworld.com/what-can-pigs-not-eat/

[43] Note that both of these references below are from Australia where the laws are different from Europe and the US. Nevertheless, they show documentation of diseases spreading through meat being fed to livestock.

https://www.agric.wa.gov.au/livestock-biosecurity/pig-feed-what-you-can-and-can%E2%80%99t-feed-pigs-swill
https://www.agric.wa.gov.au/livestock-biosecurity/pig-feed-what-you-can-and-can%E2%80%99t-feed-pigs-swill?page=0%2C0#smartpaging_toc_p0_s0_h3

[44] https://www.iamcountryside.com/pigs/what-not-to-feed-pigs/
This farmer says that in the winter 7% of the pig's diet is dairy and eggs. See the response of Walter Jefferies in this thread: https://permies.com/t/51244/Kunekune-pigs-feed

45 https://opensanctuary.org/article/things-that-are-toxic-to-pigs/

[46] https://fullboarfarm.com/2018/03/30/289/

[47] *https://www.thepigsite.com/articles/common-causes-of-poisoning-in-pigs*

[48] *https://www.triplepundit.com/story/2016/reducing-food-waste-feeding-scraps-pigs/27131*

[49] https://extension.psu.edu/raising-small-groups-of-pigs

[50] *https://sustainablefoodtrust.org/articles/food-to-pork-filling-pig-bellies-not-bins/*

https://morningchores.com/about-raising-pigs/

[51] *https://feedbackglobal.org/campaigns/pig-idea/*

[52] https://www.chlpi.org/wp-content/uploads/2013/12/Leftovers-for-Livestock_A-Legal-Guide_August-2016.pdf

[53] https://opensanctuary.org/article/things-that-are-toxic-to-pigs/

More details for mycotoxin poisoning https://www.thepigsite.com/articles/common-causes-of-poisoning-in-pigs

[54] Ibid.

[55] https://www.hobbyfarms.com/pigs-feeding-kitchen-scraps/

[56] http://www.growyourownnevada.com/pig-manure-is-it-safe-for-gardens/

[57] Ibid.

[58] https://www.motherearthnews.com/homesteading-and-livestock/raising-pigs/american-guinea-hog-zm0z15onzmat

[59] https://americankunekunepigsociety.com/

[60] https://americanKunekunepigsociety.com/breedstandard

[61] https://americanKunekunepigsociety.com/AKKPS-Color-chart-help-guide

[62] https://extension.psu.edu/animals-and-livestock/swine/swine-experts

[63] https://americanminipigassociation.com/owners/helpful-owner-articles/vaccinations/

[64] https://www.thepigsite.com/disease-guide/leptospirosis-leptospira

[65] https://www.cdc.gov/leptospirosis/symptoms/index.html

[66] https://www.thepigsite.com/disease-guide/actinobacillus-pleuropneumonia-app

[67] https://rossmillfarm.com/2019/08/common-pig-ailments/

[68] Still shot from this instructional video on hoof trimming:
https://www.youtube.com/watch?v=AbA9idAsYm4

[69]https://www.americanminipigassociation.com/

[70] https://www.wattagnet.com/articles/24486-a-practical-sow-hoof-trimming-guide

[71] https://www.youtube.com/watch?v=AbA9idAsYm4

[72] https://www.youtube.com/watch?v=AbA9idAsYm4

[73] https://www.youtube.com/watch?v=Y_kM60xRclE

[74] http://theminipigfarrier.com/2019/09/mini-pig-tusk-trimming

[75] Mozzachio K, Pollock C. Basic information sheet: miniature pig. LafeberVet Web site. July 9, 2019. Available at https://lafeber.com/vet/basic-information-sheet-miniature-pig/

[76] https://www.youtube.com/watch?v=lbK3eBX0Bz4

[77] https://americanminipigassociation.com/educational/tusk-trimming-mini-pigs-using-gigli-wire-saw/

[78] https://hoghavenblog.org/2015/06/introducing-a-new-pig-to-your-herd/

[79] https://smallfarms.cornell.edu/2016/01/pigs-n-trees/

[80] https://www.fs.usda.gov/nac/practices/silvopasture.php
"*Well-managed silvopastures* employ agronomic principals, typically including introduced or native pasture grasses, fertilization and nitrogen-fixing legumes, and rotational grazing systems that employ short grazing periods that maximize vegetative plant growth and harvest. The annual grazing income helps cash flow the tree operation while the tree crop matures and creates easy access if and when the trees or tree products are harvested. While these systems can require a number of management activities, the benefits can make it worthwhile."

[81] https://www.fs.usda.gov/nac/practices/silvopasture.php

[82] https://www.ers.usda.gov/data-products/major-land-uses/glossary.aspx

[83] https://www.youtube.com/watch?v=0zWQnc5C1Zg

[84] https://www.youtube.com/watch?v=XPIhxzHHNNg

[85] https://www.youtube.com/watch?v=o53nJsHvmVo

[86] https://www.youtube.com/watch?v=o53nJsHvmVo

[87] How I Got My Pigs to Plant Grass (in the forest)

[88] https://www.merriam-webster.com/dictionary/gley

[89] https://permies.com/t/3409/Gley-technique-sealing-ponds-dams

[90] https://farmhacker.com/seal-a-pond-naturally-with-pigs/

[91] https://www.youtube.com/watch?v=25QAbi-TAbY

[92] https://theozarkhouse.wordpress.com/2013/01/30/sealing-a-pond-with-pigs/

[93] https://theozarkhouse.wordpress.com/2013/01/30/sealing-a-pond-with-pigs/

[94] " I *was once told by a county extension agent that to seal a new pond first put down a layer of bentonite. How thick I don't know. Then, before adding any water, start feeding some hay to some cattle in the pond. Scatter it around over a period of time, letting the cows walk around in there eating the hay. I'm not sure how long to do this, but eventually they will mash the waste hay down into the bentonite and really pack it down. I don't see why the same thing wouldn't work using pigs. Except that maybe they will root around and dig up the bentonite, making thin spots, so that it doesn't make a uniform layer on the bottom.*" https://www.homesteadingtoday.com/threads/using-pigs-to-seal-a-pond.521493/

[95] Hall, Ian R.; Gordon Brown; Alessandra Zambonelli. <u>Taming the truffle: the history, lore, and science of the ultimate mushroom.</u> Timber Press: 2007.

https://en.wikipedia.org/wiki/Truffle_hog

[96] https://www.napatrufflefestival.com/truffle-hunters-friend-dog-pig/

[97] https://fdocuments.in/reader/full/how-to-train-your-truffle-pig

[98] Liles, Emma. <u>How to Train your Truffle Pig</u>. https://fdocuments.in/reader/full/how-to-train-your-truffle-pig

[99] Black, White and English truffle oil

https://www.amazon.com/TruffleHunter-Truffle-Oil-Selection-3-38/dp/B009HRQ47U/ref=sr_1_9?dchild=1&keywords=truffle+oil&qid=1620680622&sr=8-9

[100] https://www.napatrufflefestival.com/truffle-hunters-friend-dog-pig/

[101] Ibid.

[102] Here are videos of truffle hunting pigs and their trainers in action:

https://www.youtube.com/watch?v=_1L-uev-hwg

https://www.youtube.com/watch?v=hKyJ_KFgNwc

If you are interested in gourmet French cuisine, there are tours that focus on truffles and the practice of truffle hunting

https://www.exclusive-france-tours.com/truffle-season-france-private-tours-authentic-tastings/

https://www.tripadvisor.com/Attraction_Review-g608804-d3387550-Reviews-Truffle_Hunting_at_Les_Pastras-Cadenet_Luberon_Vaucluse_Provence_Alpes_Cote_d_Azu.html

[103] Image on left of girl
https://sowegalive.com/wp-content/uploads/2015/06/walking-pig.jpg

Image on right of boy
https://www.uswhip.com/blog/how-to-select-and-use-the-right-show-pig-whip/

[104] Blaine Rodgers, Show Livestock Business Development at BioZyme® Inc https://surechamp.com/picking-the-perfect-show-pig-prospect/

[105] https://americanminipigassociation.com/mini-pig-breeds/american-guinea-hog-breed-standard/

[106] https://americanminipigassociation.com/mini-pig-breeds/american-guinea-hog-breed-standard/

[107] https://www.americanKunekunepigregistry.com/sanctioned-show-rules

[108] https://www.americanKunekunepigregistry.com/sanctioned-show-rules

[109] https://americanKunekunepigsociety.com/AKKPS-Color-chart-help-guide

[110] https://www.youtube.com/watch?v=KdSQs7jbg7E

[111] This is a very useful .pdf about the care and feeding of show pigs from the Utah State University Extension.
https://www.showpig.com/EDUCATION/BasicShowPigFeedingandCare.pdf

[112] https://www.uswhip.com/blog/how-to-raise-a-winning-show-pig/

[113]
https://www.pewtrusts.org/~/media/assets/2017/07/alternatives_to_antibiotics_in_animal_agriculture.pdf

[114] This .pdf from the Utah State University Extension gives the details of the nutritional needs for show pigs
https://www.showpig.com/EDUCATION/BasicShowPigFeedingandCare.pdf

This supplier offers feed for the various stages of life as well as extra minerals and supplements
https://www.showrite.com/pig/

[115] https://www.uswhip.com/blog/how-to-raise-a-winning-show-pig/

116 https://www.hogslat.com/farmstead-single-door-hog-pig-feeder-fs400?msclkid=63462dc432bd1238ebe14d5aa5241802&utm_source=bing&utm_medium=cpc&utm_campaign=**LP%20-%20Shop2%20-%20Feeding&utm_term=4575617641213112&utm_content=FS400%20%7C%20FARMSTEAD%20Single%20Door%20Pig%20Feeder%20%7C%20%2496.49%20USD

117 https://www.showpig.com/EDUCATION/BasicShowPigFeedingandCare.pdf

118 https://www.youtube.com/watch?v=ehoU6Kq6gAo

119 https://www.showstopperequipment.com/show-pig-whips

120 https://www.uswhip.com/blog/how-to-select-and-use-the-right-show-pig-whip/

121 This is an excellent place to start to understand how the whip is used.
https://www.uswhip.com/blog/how-to-select-and-use-the-right-show-pig-whip/

This article has close up photos of a hand demonstrating how to hold the pig whip as well as a video to demonstrate the use of it to guide the pig.
https://surechamp.com/how-do-your-use-your-show-pig-whip/

122 https://www.uswhip.com/blog/how-to-keep-your-show-pig-from-running/

123 https://agrilifeextension.tamu.edu/library/4-h-youth-development/keeping-show-pigs-healthy/

124 https://agrilifeextension.tamu.edu/library/4-h-youth-development/keeping-show-pigs-healthy/

125 https://agrilifeextension.tamu.edu/library/4-h-youth-development/keeping-show-pigs-healthy/

126 Ibid.

127 https://www.cdc.gov/flu/swineflu/exhibit-pigs-at-fairs.htm

128 https://www.cdc.gov/flu/swineflu/exhibit-pigs-at-fairs.htm

129 http://coloradomountaindogs.com/lwcmds

130 Canine Pack Partners for Karakachans
https://dogfolksdirtfarm.com/why-we-breed

Colorado Mountain Dogs
http://coloradomountaindogs.com/

131 **Livestock Guardian Donkeys**

Pros and Cons
https://www.motherearthnews.com/homesteading-and-livestock/guard-donkey-zbcz1310

Ontario Ministry of Agriculture articles about donkey guardians
http://www.omafra.gov.on.ca/english/livestock/sheep/facts/donkey2.htm

Modern Farmer's Guide to Guard Donkeys
https://modernfarmer.com/2014/06/modern-farmers-guide-guard-donkeys/

Livestock Guardian Llamas
Pros and Cons
https://www.motherearthnews.com/homesteading-and-livestock/guardian-llamas-zbcz1309

[132] https://extension.psu.edu/raising-small-groups-of-pigs

[133] https://guineahogs.org/life-cycle-of-american-guinea-hogs/

[134] https://guineahogs.org/life-cycle-of-american-guinea-hogs/

[135] https://www.Kunekunebreeders.com/Kunekune-Pigs-What-is-COI.html

Also, the breed associations can give you resources about line breeding for desirable traits vs inbreeding when the COI is too high.

This article is very useful about breeding Kunekunes generally
https://www.virginiaKunekunes.com/Breeding%20and%20Farrowing.pdf

American Guinea Hog Association
https://guineahogs.org/life-cycle-of-american-guinea-hogs/

American Kunekune Pig Society
https://americanKunekunepigsociety.com/

The New Zealand Kunekune Pig Association has this resource on breeding
http://Kunekune.co.nz/index.php/Kunekune-breeding/

[136] http://Kunekune.co.nz/index.php/Kunekune-breeding/

[137] https://extension.wsu.edu/animalag/content/environmental-management-of-young-pigs-during-cool-weather/

[138] https://guineahogs.org/life-cycle-of-american-guinea-hogs/

[139] http://board.britishKunekunesociety.org.uk/viewtopic.php?t=1290

[140] https://www.hsa.org.uk/humane-killing-of-livestock-using-firearms-positioning/pigs-2

[141] Improved handling systems for pigs at slaughter
https://www.hsa.org.uk/shop/publications-1/product/improved-handling-systems-for-pigs-at-slaughter

Human Slaughter of Pigs
https://www.hsa.org.uk/shop/publications-1/product/humane-slaughter-of-pigs

[142] https://extension.psu.edu/raising-small-groups-of-pigs

[143] 2:30 is where the lesson begins with the washing of the pig. I have a dedicated scrub brush for this purpose, I really like the brush they are using.
https://www.youtube.com/watch?v=qdPGWkY3rL4

[144] Poking hole in foot for hanging at 4:57
https://www.youtube.com/watch?v=qdPGWkY3rL4

[145] At 6:30 he starts to tie the knots on the nails to prepare for hanging
https://www.youtube.com/watch?v=qdPGWkY3rL4

[146] 6:55 another wash down after gutting
https://www.youtube.com/watch?v=31zqAQX26GY

[147] Even though the pig is on a table, you an see that this can be done when hanging as well.
https://www.youtube.com/watch?v=ivBdHcsF83M

[148] Cutting the pig into pieces for your freezer
https://www.youtube.com/watch?v=ivBdHcsF83M

[149] https://uknowledge.uky.edu/cgi/viewcontent.cgi?article=1000&context=yield_reports